Busy Ant Maths

Test Pack 4

Series Editor: Peter Clarke

Author: Caroline Fawcus

William Collins' dream of knowledge for all began with the publication of his first book in 1819. A self-educated mill worker, he not only enriched millions of lives, but also founded a flourishing publishing house. Today, staying true to this spirit, Collins books are packed with inspiration, innovation and practical expertise. They place you at the centre of a world of possibility and give you exactly what you need to explore it.

Collins. Freedom to teach.

Published by Collins
An imprint of HarperCollins*Publishers*
The News Building
1 London Bridge Street
London
SE1 9GF

Browse the complete Collins catalogue at
www.collins.co.uk

ISBN 978-0-00-816739-4

British Library Cataloguing in Publication Data
A Catalogue record for this publication is available from the British Library.

Author: Caroline Fawcus
Series Editor: Peter Clarke
Commissioning Editor: Fiona McGlade
Project Editor: Leah Willey
Cover design and artwork: Amparo Barrera
Internal design and illustrations: Steve Evans and Mike Connor
Production: Robin Forrester

Printed and bound by CPI Group (UK) Ltd, Croydon, CR0 4YY

Contents

Introduction

Purpose of the Busy Ant Maths Year 4 Test Pack

1. To assist teachers in assessing the overall level of mastery achieved by pupils in the mathematics programme of study for Year 4, and in determining pupils' readiness to deal with the expectations of the Year 5 programme of study.

2. To provide parents, teachers and school leaders with information on how pupils are performing in comparison with national standards, by assigning one of the following:
   - working below national standard
   - working towards national standard
   - working at national standard
   - working at greater depth within the national standard.

3. To enable teachers and school leaders to monitor the performance of pupil cohorts, to identify where interventions may be required and to ensure that pupils are supported to achieve sufficient progress and expected attainment.

The Busy Ant Maths Test Packs are intended to be used as summative assessments towards the end of the academic year. They are not designed to assess pupils' level of mastery in each of the individual National Curriculum attainment targets and domains. For guidance on making formative assessments regarding pupils' strengths and weaknesses and level of mastery in specific attainment targets and domains, teachers should refer to the Assessment Tasks and Assessment Exercises in the Busy Ant Maths Assessment Guides.

Structure of the Busy Ant Maths Year 4 Test Pack

Paper	Paper 1: arithmetic	Paper 2: reasoning	Paper 3: reasoning
Total marks	30 marks	30 marks	30 marks
Recommended timing	25 minutes	30 minutes	30 minutes
National Curriculum domain coverage	Number – Number and place value Number – Addition and subtraction Number – Multiplication and division Number – Fractions (including decimals)		
		Measurement Geometry – Properties of shapes Geometry – Position and direction Statistics	

The papers should be administered in order.

Pupils can have a break between the papers.

Administering Paper 1: arithmetic

Recommended timing

Pupils have 25 minutes to complete the paper.

Resources

Pupils will need the following resources:

- a blue or black pen, or a dark pencil
- a rubber (optional). If rubbers are not provided, tell pupils that they may cross out any answer they wish to change.

Pupils may also use the following resources, if this is normal classroom practice, provided they only give word-for-word translations:

- bilingual dictionaries or electronic translators
- bilingual word lists
- monolingual English electronic spell checkers.

Pupils are not allowed:

- calculators
- other mathematical equipment, such as angle measurers, mirrors or rulers.

Assistance

Teachers should ensure that nothing they do or say during the test could be interpreted as giving pupils an advantage.

If a pupil requests it, teachers may read a question out loud on a one-to-one basis.

If reading to a pupil, teachers should read words and numbers but not mathematical symbols. This is to ensure that pupils are not given an unfair advantage by having the operation/function inadvertently explained by reading its name.

Before the test begins

- Review the list of pupils with any particular individual needs and ensure that their requirements are met.
- Ensure that pupils are able to work undisturbed and that the classroom layout and seating arrangements are suitable.
- Ensure that any wall displays do not give pupils an unfair advantage.
- Write the start and finish times of the test on the board.
- Provide each pupil with the resources they require.

What to say at the start of the test

- *This is the arithmetic paper.*
- *You will need a blue or black pen or a dark pencil. You may use a rubber for this test.* (If rubbers are not provided, tell the pupils that they may cross out any answer that they wish to change.)
- *Write your name and class name/number on the front of your test paper.*
- *Look at the list of instructions on page 1. I will read the instructions to you.* (Read the instructions on page 1 of the test paper to the pupils.)

Instructions

You **may not** use a calculator to answer any questions in this test.

Questions and answers

You have **25 minutes** to complete this test.

Work as quickly and as carefully as you can.

Put your answer in the box for each question.

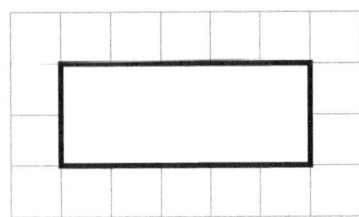

For questions expressed as common fractions, you should give your answer as common fractions.

All other answers should be given as whole numbers or decimals.

If you cannot do one of the questions, **go on to the next one**. You can come back to it later if you have time.

If you finish before the end, **go back and check your work**.

Marks

The number under each box at the side of the page tells you the maximum number of marks for each question.

In this test, some multiplication questions are worth **2 marks each**. You will be awarded 2 marks for a correct answer. You may get 1 mark for showing the formal method.

All other questions are worth **1 mark each**.

- *If you want to change your answer, put a line through the response you don't want, or use a rubber.*
- *Remember to check your work carefully.*
- *If you have any questions during the test, put your hand up and wait for someone to come to you. Remember, I can't help you answer any of the test questions.*
- *You must work on your own and you must not talk to each other.*
- *Are there any questions you want to ask me now?*
- *I will tell you when you have 5 minutes left.*
- *I will tell you when the test is over and to stop writing.*
- *You may now start the test.*

Administering Papers 2 and 3: reasoning

Recommended timing

Pupils have 30 minutes to complete each paper.

Resources

Pupils will need the following resources for each test:

- a blue or black pen, or a dark pencil
- a sharp, dark pencil for mathematical drawing
- a ruler (showing centimetres and millimetres) (Paper 3: reasoning only)
- a rubber (optional). If rubbers are not provided, tell the pupils that they may cross out any answer that they wish to change.

Pupils may also use the following resources, if this is normal classroom practice, provided they only give word-for-word translations:

- bilingual dictionaries or electronic translators
- bilingual word lists
- monolingual English electronic spell checkers.

Pupils are not allowed:

- calculators.

Assistance

Teachers should ensure that nothing they do or say during the tests could be interpreted as giving pupils an advantage.

If a pupil requests it, teachers may read a question out loud on a one-to-one basis.

If reading to a pupil, teachers should read words and numbers but not mathematical symbols. This is to ensure that pupils are not given an unfair advantage by having the operation/function inadvertently explained by reading its name.

At a pupil's request, teachers can point to parts of the test paper such as charts, diagrams, statements and equations, but they should not explain the information or help the pupil by interpreting it.

If any everyday context or words related to a question are unfamiliar to a pupil, teachers may show them related objects or pictures, or describe the related context.

Before each test begins

- Review the list of pupils with any particular individual needs and ensure that their requirements are met.
- Ensure that pupils are able to work undisturbed and that the classroom layout and seating arrangements are suitable.
- Ensure that any wall displays do not give pupils an unfair advantage.
- Write the start and finish times of the test on the board.
- Provide each pupil with the resources they require.

What to say at the start of the test

- *This is the first/second reasoning paper.*
- *You will need a blue or black pen or dark pencil, a sharp pencil for mathematical drawing, and a ruler (Paper 3: reasoning only). You may use a rubber for this test.* (If rubbers are not provided, you should tell the pupils that they may cross out any answer that they wish to change.)
- *Write your name and class name/number on the front of your test paper.*
- *Look at the list of instructions on page 1. I will read the instructions to you.* (Read the instructions on page 1 of the test paper to the pupils.)

Instructions

You **may not** use a calculator to answer any questions in this test.

Questions and answers

You have **30 minutes** to complete this test.

Follow the instructions for each question.

Work as quickly and as carefully as you can.

If you need to do working out, you can use the space around the question.

Some questions have a method box like this:

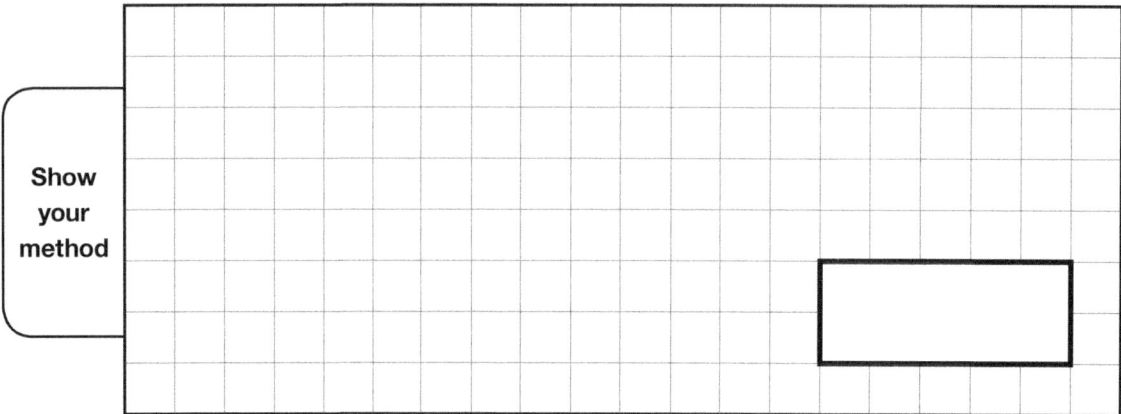

For these questions you may get a mark for showing your method.

If you cannot do one of the questions, **go to the next one**.
You can come back to it later, if you have time.

If you finish before the end, **go back and check your work**.

Marks

The number under each line at the side of the page tells you the maximum number
of marks for each question.

- *If you want to change your answer, put a line through the response you don't want, or use a rubber.*
- *Remember to check your work carefully.*
- *If you have any questions during the test, put your hand up and wait for someone to come to you.*
  Remember, I can't help you answer any of the test questions.
- *You must work on your own and you must not talk to each other.*
- *Are there any questions you want to ask me now?*

- *I will tell you when you have 5 minutes left.*
- *I will tell you when the test is over and to stop writing.*
- *You may now start the test.*

What to do at the end of the test

- Use the mark schemes to mark the two papers, following the '**General marking guidance**', the '**Requirements**', and any '**Additional guidance**' for each question.
- Write the total score for each paper on the front of the paper.
- Calculate the total marks allocated for all three papers to determine the pupil's overall score out of 90. Record their results.
- If you wish to compare pupils' achievements with national standards, refer to the **Assigning a national standard** table at the end of this Introduction.

Explanation of the mark schemes

The marking information for each question is set out in the form of tables which start on page xiii.

The '**Qu.**' column on the left-hand side of each table provides a quick reference to the question number and, where applicable, the question part.

The '**Requirement**' column may include two types of information:

- a statement of the requirements for the award of each mark, with an indication of whether credit can be given for a correct method
- examples of some different types of correct response.

The '**Mark**' column indicates the total number of marks available for each question part.

The '**Additional guidance**' column indicates alternative acceptable responses, and provides details of specific types of response that are unacceptable. Other guidance, such as the range of acceptable answers, is provided as necessary.

General marking guidance

What if...	Marking procedure
The pupil's response is numerically equivalent to the answer in the mark scheme.	Award the mark unless the mark scheme states otherwise.
The pupil's response does not match closely any of the examples given.	Teachers should use their professional judgement in deciding whether the response corresponds with the statement of the requirements given in the 'Requirement' column. Consideration should also be given to the 'Additional guidance' column.
The pupil has responded in a non-standard way.	Pupils may provide evidence in a form as long as its meaning can be understood. Diagrams, symbols or words are acceptable for explanations or for indicating a response. In Paper 1: arithmetic, pupils should use formal methods for calculating their answers. For those questions that are worth 2 marks, a partial credit of 1 mark will be awarded for evidence of using formal methods with one arithmetical error. In Papers 2 and 3: reasoning, a partial mark (or marks) will be awarded for evidence of a complete and correct method.
There appears to be a misreading affecting the working.	If the original intention or difficulty level of the question is not reduced deduct 1 mark only. In 1-mark questions, 0 marks are awarded. In 2-mark questions that have a method mark, 1 mark should be awarded if the correct method is correctly implemented with the misread number. If the original intention or difficulty level of the question is reduced do not award any mark for the question part.
No answer is given in the expected place, but the correct answer is given elsewhere.	Where a pupil has shown understanding of the question, the mark(s) should be given.
The pupil's answer is correct but the wrong working is shown.	A correct response should always be marked as correct unless the mark scheme states otherwise.
The pupil has worked out the answer correctly and then written an incorrect answer in the answer box.	Precedence should be given to the answer given in the answer box over any other workings. There may be cases where the incorrect answer is due to a transcription error. In such cases check the pupil's intention and decide whether to award the mark.
The correct response has been crossed (or rubbed) out and not replaced.	Any legible crossed-out work that has not been replaced should be marked according to the mark schemes. If the work is replaced, then crossed-out work should not be considered.
More than one answer is given.	If all answers are correct (or a range of answers is given, all of which are correct), the mark should be awarded unless prohibited by the mark schemes. If both correct and incorrect responses are given, no mark should be awarded unless the mark schemes state otherwise.
The answer is correct but, in a later part of the question, the pupil has contradicted this response.	A mark given for one part should not be disallowed for working or answers given in a different part, unless the mark scheme specifically states otherwise.

Responses involving money

	Accept	Do not accept
Where the £ is given: £	£1.60 £5 £5.00 Any unambiguous indication of the correct amount, e.g. £1.60p £1 60 pence £1 60 £1,60 £1-60 £1:60	Incorrect placement of pounds or pence, e.g. £160 £160p Incorrect placement of decimal point, or incorrect use or omission of 0, e.g. £1.6 £1 600 £16 0 £1-6-0
Where the p is given: p	20p Any unambiguous indication of the correct amount, e.g. £0.20p	Incorrect or ambiguous use of pounds or pence, e.g. 0.20p £20p
Where no sign is given:	£1.60 20p 160p £0.20 Any unambiguous indication of the correct amount, e.g. £1.60p £0.20p £1 60 pence £.20p £1 60 £.20 £1,60 20 £1-60 0.20 £1:60 1.60 160 1 pound 60	Incorrect or ambiguous use of pounds or pence, e.g. £160 £20 £160p £20p £1.6 0.2 1.60p 0.20p

Responses involving time

	Accept	Do not accept
A time interval, e.g. 1 hour 30 minutes	1 hour 30 minutes Any unambiguous, correct indication, e.g. $1\frac{1}{2}$ hours 1·5 hours 1h 30 1h 30 min 1 30 90 minutes 90 Digital electronic time, i.e. 1:30	Incorrect or ambiguous time interval, e.g. 1.30 1–30 1,30 130 1·3 1·3 hours 1·3h 1h 3 1·30 min
A specific time, e.g. 3:20 a.m., 15:20	3:20 a.m. 3:20 twenty past three Any unambiguous, correct indication, e.g. 03·20 3·20 0320 3 20 3-20 3,20 Unambiguous change to 12- or 24-hour clock, e.g.: 18:10 as 6:10 p.m. or 18:10 p.m.	Incorrect time, e.g. 3·2 a.m. 3·20 p.m. Incorrect placement of separators, spaces, etc. or incorrect use or omission of 0, e.g. 320 3:2:0 3·2 032

Responses involving measures

	Accept	Do not accept
Where units are given, e.g. m, kg, l ☐ m	1·5 m Any unambiguous indication of the correct measurement, e.g. $1\frac{1}{2}$ m 1·50 m 1 m 50 cm	Incorrect or ambiguous use of units, e.g. 1500 m

Notes:

- If a pupil leaves the answer box empty but writes the answer elsewhere on the page, then that answer must be consistent with the units given in the answer box and the conditions listed above.
- If a pupil changes the unit given in the answer box, then their answer must be equivalent to the correct answer using the unit they have chosen, unless otherwise indicated in the mark schemes.

Mark schemes for Paper 1: arithmetic

Qu.	Requirement	Mark	Additional guidance
1	48	1	
2	7·7	1	Accept equivalent fractions, e.g. $7\frac{7}{10}$
3	1078	1	
4	$\frac{2}{11}$	1	Accept equivalent fractions.
5	2199	1	
6	5	1	
7	200	1	
8	4	1	
9	4804	1	
10	12	1	
11	$1\frac{1}{3}$	1	Accept improper fraction, i.e. $\frac{4}{3}$
12	61·3	1	
13	84	1	
14	7·73	1	
15	1120	1	
16	215	1	
17	0·8	1	Accept equivalent fractions, e.g. $\frac{8}{10}$
18	3181	1	
19	$\frac{5}{6}$	1	
20	110	1	
21	£4.05	1	

Qu.	Requirement	Mark	Additional guidance
22	Award **TWO** marks for the correct answer of 208 If the answer is incorrect, award **ONE** mark for the formal multiplication method which contains no more than **ONE** arithmetical error, e.g. $$\begin{array}{r} 2\ 6 \\ \times\quad 8 \\ \hline \text{Wrong answer} \\ \hline 4 \end{array}$$	Up to 2	**Do not** award any marks if there is no evidence of the ones (i.e. their 4) being carried into the tens column.
23	921	1	
24	5820	1	
25	60	1	
26	Award **TWO** marks for the correct answer of 1268 If the answer is incorrect, award **ONE** mark for the formal multiplication method which contains no more than **ONE** arithmetical error, e.g. $$\begin{array}{r} 3\ 1\ 7 \\ \times\quad 4 \\ \hline \text{Wrong answer} \\ \hline 2 \end{array}$$	Up to 2	**Do not** award any marks if there is no evidence of the ones (i.e. their 2) being carried into the tens column.
27	0·07	1	Accept equivalent fractions, e.g. $\frac{7}{100}$
28	22·19	1	

Mark schemes for Paper 2: reasoning

Qu.	Requirement	Mark	Additional guidance
1	Three numbers circled, largest in each line, as shown. (7800) 7080 4545 (4554) (9191) 9119	1	All three answers must be correct for the award of the mark. Part marks are not awarded.
2	Three lines correctly drawn, as shown. 540 545 560 450 460 440 550 455	1	**Do not** award the mark if the answer given is not a one-to-one mapping.
3	A two-part instruction, given in either order, of: RIGHT 1 **AND** UP 3	1	**Do not** accept instructions of more than 2 parts, e.g. RIGHT 2, LEFT 1, UP 3 **Do not** accept arrows (→ ↑) instead of Right and Up.
4	−4 °C	1	**Do not** award the mark for 4− °C
5		1	

Qu.	Requirement	Mark	Additional guidance
6		1	
7		1	Pupils are asked in the question to circle their answers, although any correct indication of the mark, e.g. underlining or arrows, providing it is unambiguous will be awarded the mark.
8	Award **ONE** mark for each correct answer of 3·25 and 1·75 in the correct position in the table.	Up to 2	
9	Award **TWO** marks for the correct answer of 2369 If the answer is incorrect, award **ONE** mark for evidence of a correct method, e.g. 6392 + 7239 = 13 631 16 000 − 13 631 = OR 8000 − 6392 = 1608 8000 − 7239 = 761 761 + 1608 =	Up to 2	A final answer does not need to be given for the award of **ONE** mark. However the method must be complete, so an answer that calculated 1608 and 761 respectively without further calculations would not gain any credit.
10	$6\frac{1}{5}$	1	Accept equivalent fractions and decimals. e.g. 6·2
11	9	1	
12		1	**Do not** award the mark if the answer given is not a one-to-one mapping.
13a 13b	12 	1 1	Accept inaccurate drawings if intention is clear.

Qu.	Requirement	Mark	Additional guidance
14		1	
15a		1	
15b	0·24	1	
16	Award **ONE** mark for each correct answer in the correct position of the table, as shown:	Up to 2	
17	Award **TWO** marks for the correct answer of £52.80 If the answer is incorrect, award **ONE** mark for evidence of a correct method, e.g. £4.80 × 10 = £48.00 £2.40 × 2 = £4.80 £48.00 + £4.80 =	Up to 2	
18	Two shapes with an area of 9 units^2 ticked, as shown:	1	
19	Award **TWO** marks for the correct answer of 2 If the answer is incorrect, award **ONE** mark for evidence of a correct method, e.g. 7 × 12 = 84 86 – 84 = OR 7 × 13 = 91 91 – 86 = 5 7 – 5 =	Up to 2	

Qu.	Requirement	Mark	Additional guidance
20		1	
21	Award **TWO** marks for a correct answer of 296 **AND** evidence that the table has been used, e.g. 148 × 2 = 296 Award **ONE** mark for a correct use of the table resulting in an incorrect answer, e.g. 148 × 2 = incorrect answer	Up to 2	If no method or an incorrect method is shown, award **ONE** mark for a correct answer of 296
22		1	Accept the numbers written in any order as long as the numbers on each line of the cross total 160

Mark schemes for Paper 3: reasoning

Qu.	Requirement	Mark	Additional guidance
1	7,100 > 3,750 ticked, as shown: 6110 < 3705 ☐ 7100 > 3705 ☐ 6110 < 3750 ☐ 7100 > 3750 ✔	1	
2	Two flags with only one line of symmetry ticked, as shown: ☐ ✔ ✔ ☐	1	**Do not** award the mark if extra flags are ticked.
3	3163 **AND** 6163	1	Answers must be given in the correct order.
4	2	1	Accept answer indicated on the grid, e.g. mouse 2 circled.
5	10 **AND** 650	1	Both answers must be correct for the award of the mark. Answers must be given in the correct order.
6	Award **TWO** marks for a complete correct answer, as shown: 	×	3	7	9
2	6	14	18
7	21	49	63
11	33	77	99 | If the answer is incomplete or incorrect, award **ONE** mark for any three correct numbers.	Up to 2	
7	An arrow drawn pointing to 40, as shown: 0 10 20 30 40 50 60 70 80 90 100	1	

Qu.	Requirement	Mark	Additional guidance
8	(3, 6)	1	
9	Award **TWO** marks for the correct answer of £85 If the answer is incorrect, award **ONE** mark for evidence of an appropriate method, e.g. 143 × 5 = 715 800 − 715 =	Up to 2	
10	Any three numbers that correctly multiply to 100, e.g. 2 × 5 × 10 10 × 10 × 1 25 × 2 × 2 100 × 2 × $\frac{1}{2}$	1	**Do not** award the mark if any box is left blank, e.g. 10 × 10 × ☐ = 100 Award the mark for correct answers containing fractions and/or negative numbers.
11	Award **TWO** marks for correctly circling the largest in all four pairs, as shown: £5.64 — (814 p) 6·3 cm — (85 mm) (24 m) — 960 cm 1800 ml — (2 litres) If the answer is incomplete or incorrect, award **ONE** mark for any **TWO** correct.	Up to 2	
12a 12b	10 °C An answer that identifies 6 a.m. AND 4 p.m.	1 1	
13	Correct answer ticked, as shown: ☐ All quadrilaterals have two obtuse angles and 2 acute angles. ✔ All quadrilaterals have four sides. ☐ All quadrilaterals have one or more lines of symmetry. ☐ All quadrilaterals have one pair of parallel sides.	1	**Do not** award the mark if more than one box is ticked.
14	Any number between 6·46 and 6·64, e.g. 6·5 6·47 6·635	1	Answers can be given to any number of decimal places. Accept answers given as a fraction, e.g. $6\frac{1}{2}$
15	20 cm	1	

Qu.	Requirement	Mark	Additional guidance
16	48 circled, as shown: 	1	**Do not** award the mark if additional numbers are circled.
17	4·3	1	
18	15	1	
19	Award **ONE** mark for each correct equivalent fraction, e.g. $\frac{2}{8}$, $\frac{3}{12}$, $\frac{40}{160}$	Up to 2	
20	A 5 × 3 rectangle in either orientation, e.g. 	1	A rectangle of 5 × 3 is the only size that will fit on the grid provided in the question. If pupils accurately extend the grid to enable them to create a 15 × 1 rectangle then credit will be given.
21		1	
22	Award **ONE** mark for each correct answer of 4375 and 4745 in either order.	Up to 2	
23	Award **ONE** mark for each correct answer of 132 and 44, given in the correct place on the diagram, as shown: 	Up to 2	

Assigning a national standard

A total of 90 marks are available for the Year 4 Test:

- Paper 1: arithmetic (30 marks)
- Paper 2: reasoning (30 marks)
- Paper 3: reasoning (30 marks)

The sum of the marks allocated from these two papers determines the pupil's overall raw score.

Schools that wish to compare their pupils' achievements with national standards should use the thresholds in the table below to convert raw scores into national standards.

Raw score (out of 90)	0–17	18–44	45–71	72–90
Scaled score	0%–19%	20%–49%	50%–79%	80%–100%
Standard	Working below national standard	Working towards national standard	Working at national standard	Working at greater depth within the national standard

Arithmetic

Name: _____

Class: _____

Instructions

You **may not** use a calculator to answer any questions in this test.

Questions and answers

You have **25 minutes** to complete this test.

Work as quickly and as carefully as you can.

Put your answer in the box for each question.

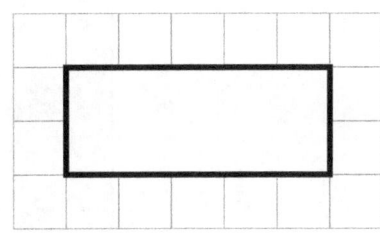

For questions expressed as common fractions, you should give your answer as common fractions.

All other answers should be given as whole numbers or decimals.

If you cannot do one of the questions, **go on to the next one**. You can come back to it later if you have time.

If you finish before the end, **go back and check your work**.

Marks

The number under each box at the side of the page tells you the maximum number of marks for each question.

In this test, some multiplication questions are worth **2 marks each**. You will be awarded 2 marks for a correct answer. You may get 1 mark for showing the formal method.

All other questions are worth **1 mark each**.

1 $8 \times 6 =$

1 mark

2 $3·6 + 4·1 =$

1 mark

3 $78 + 1000 =$

1 mark

4

$$\frac{9}{11} - \frac{7}{11} =$$

1 mark

5

2147 + 52 =

1 mark

6

30 ÷ 6 =

1 mark

7 125 + 25 + 25 + 25 =

1 mark

8 $\frac{1}{9}$ of 36 =

1 mark

9 5604 − 800 =

1 mark

10 132 ÷ 11 =

1 mark

11 $\frac{2}{3} + \frac{2}{3} =$

1 mark

12 32·6 + 28·7 =

1 mark

13

$7 \times 3 \times 4 =$

1 mark

14

$6\cdot7 + 1\cdot03 =$

1 mark

15

$756 + 364 =$

1 mark

16 43 × 5 =

1 mark

17 8 ÷ 10 =

1 mark

18 4173 − 992 =

1 mark

19

$1 - \dfrac{1}{6} =$

1 mark

20

$330 \div 3 =$

1 mark

21

£1.60 + £2.45 =

1 mark

22

Show your method

$$\begin{array}{r} 2\ 6 \\ \times\quad\ \ 8 \\ \hline \end{array}$$

2 marks

23   1000 − 79 =

1 mark

24   3106 + 2714 =

1 mark

25

$\dfrac{5}{6}$ of 72 =

1 mark

26

Show your method

```
      3 1 7
 ×        4
  _____
```

2 marks

27

7 ÷ 100 =

1 mark

28   30 – 7·81 =

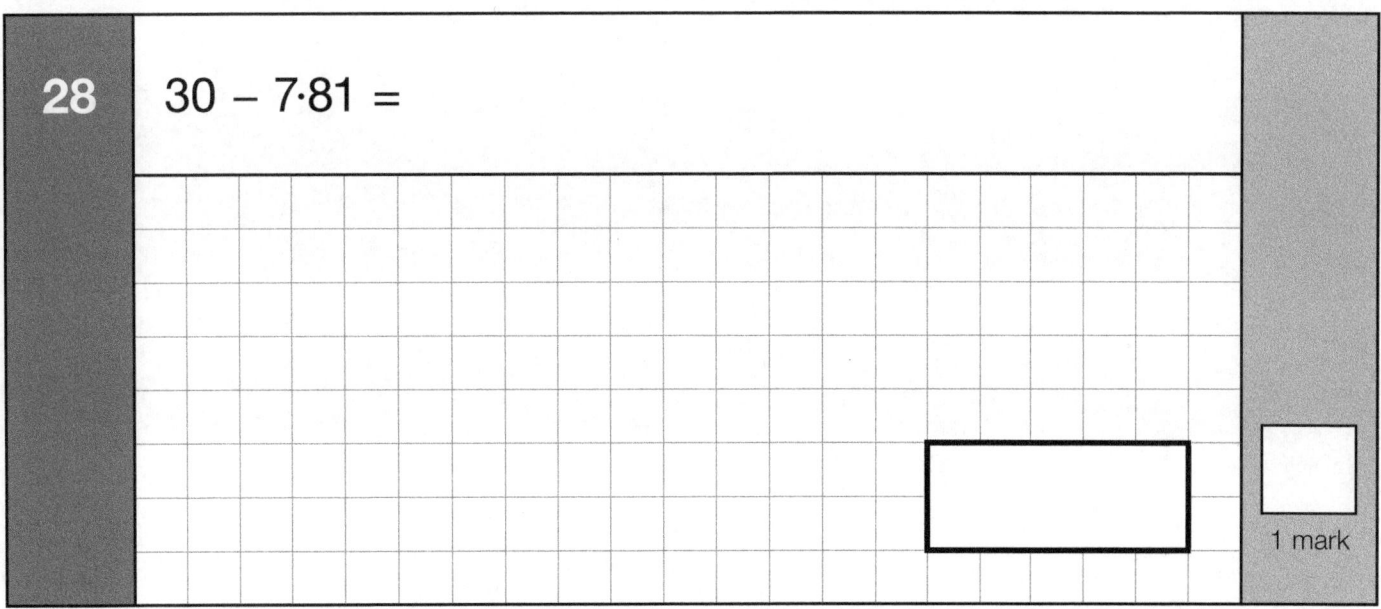

1 mark

Reasoning

Name: _____

Class: _____

Instructions

You **may not** use a calculator to answer any questions in this test.

Questions and answers

You have **30 minutes** to complete this test.

Follow the instructions for each question.

Work as quickly and as carefully as you can.

If you need to do working out, you can use the space around the question.

Some questions have a method box like this:

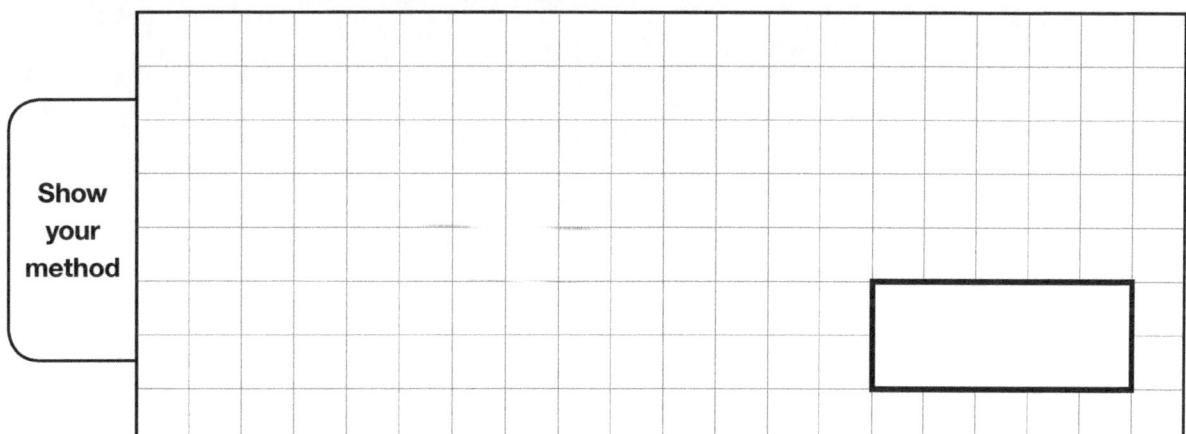

Show your method

For these questions you may get a mark for showing your method.

If you cannot do one of the questions, **go to the next one**.
You can come back to it later, if you have time.

If you finish before the end, **go back and check your work**.

Marks

The number under each line at the side of the page tells you the maximum number of marks for each question.

1 Circle the **largest** number in each pair.

One has been done for you.

3500	(5300)

7800	7080

4545	4554

9191	9119

1 mark

One has been done for you.

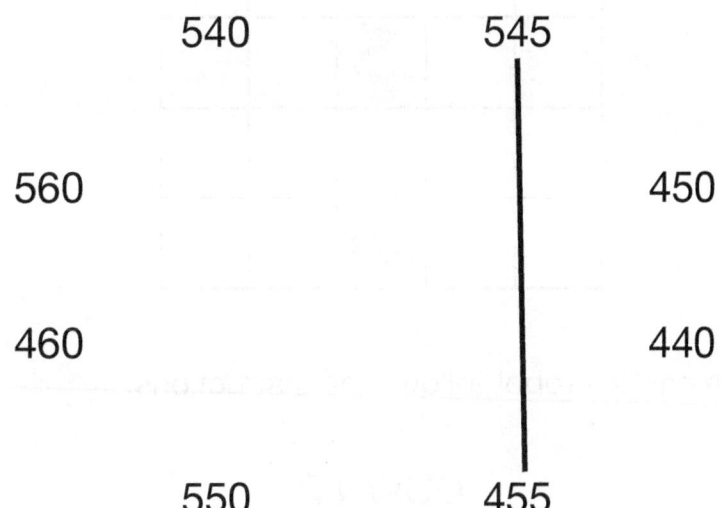

540   545

560   450

460   440

550   455

3 A robot is in the middle of the grid.

To move to the car, the robot follows the instructions:

DOWN 2

The robot is in a new position on the grid.

Write instructions to move the robot to the **tree**.

Use only **UP** or **DOWN**, and **LEFT** or **RIGHT**.

1 mark

4 The temperature is 7 degrees Celsius.

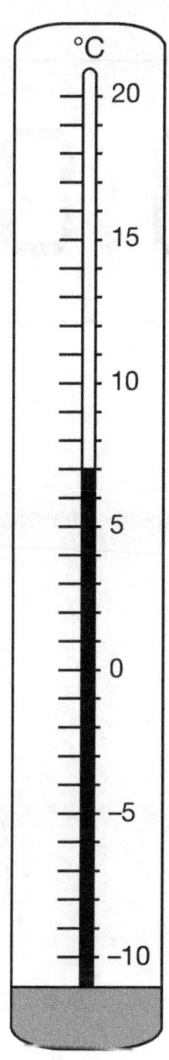

The temperature falls by 11 degrees Celsius.

What is the new temperature?

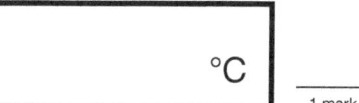

°C

1 mark

5 This is the time on a digital clock.

Tick (✔) the same time on an analogue clock.

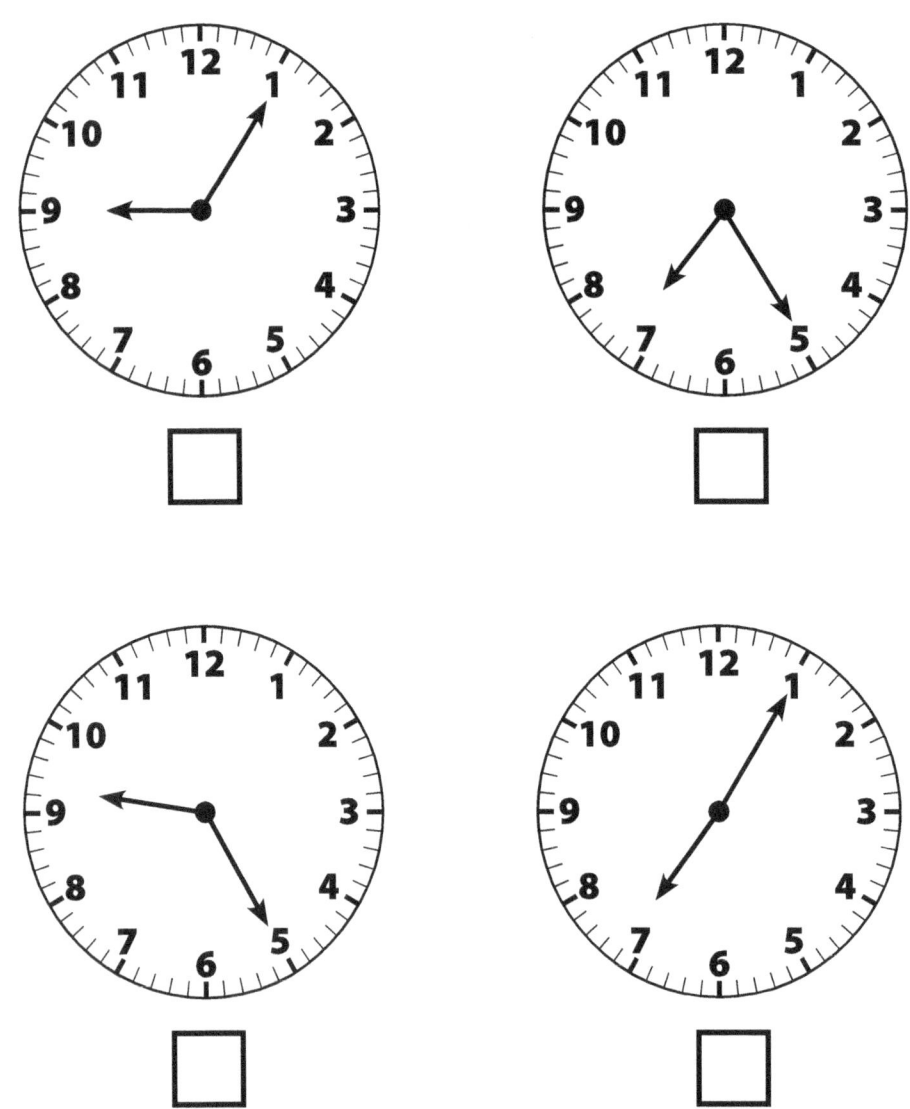

1 mark

6 Tick (✔) the shape that can be cut along one of its lines of symmetry to create two trapeziums.

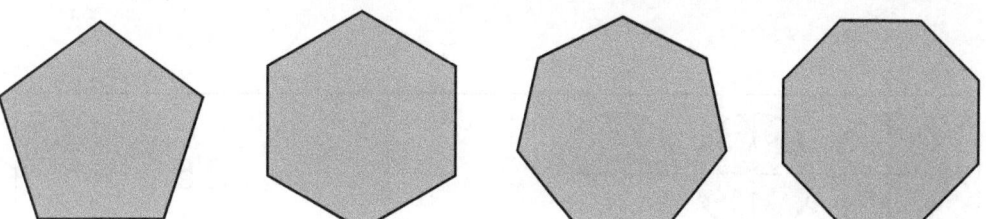

1 mark

7 Circle all the numbers that are **not** in the 7 multiplication table.

27   49   56   70   84

1 mark

8   Write the decimal equivalents shown by these shapes.

One has been done for you.

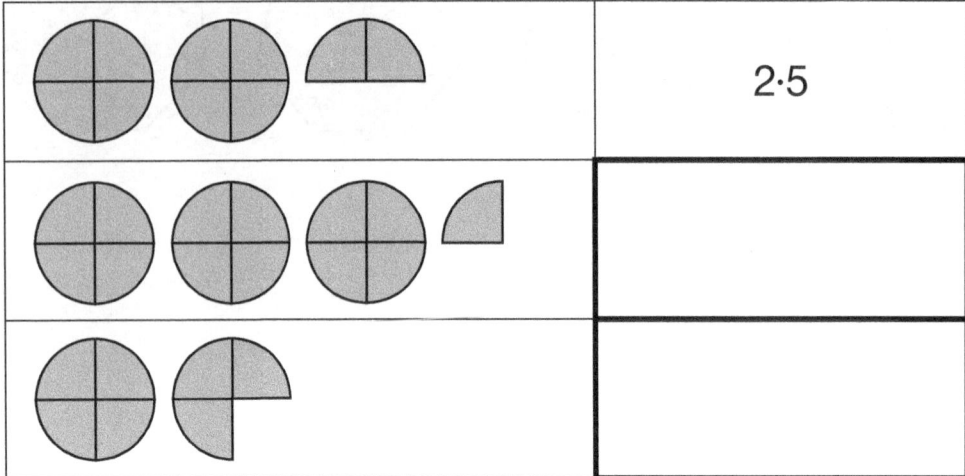

2 marks

9 A stadium can hold 8000 people when it is full.

The table below shows the number of people that attended a concert one weekend.

Day	Number of people
Saturday	6392
Sunday	7239

How many **more** people could have attended altogether?

Show your method

2 marks

10 Complete the calculation.

$$3\frac{4}{5} + \boxed{} = 10$$

1 mark

11 Round 8·7 to the nearest whole number.

1 mark

12 Four angles are marked on the diagram.

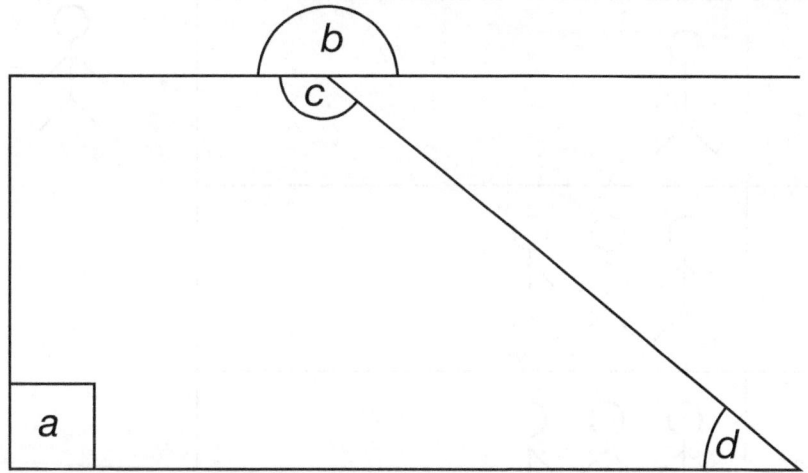

Match the letter of each angle to its name.

One has been done for you.

a			acute
b			obtuse
c			right angle
d			straight angle

1 mark

13 A teacher counted the number of children attending a club each day for a week.

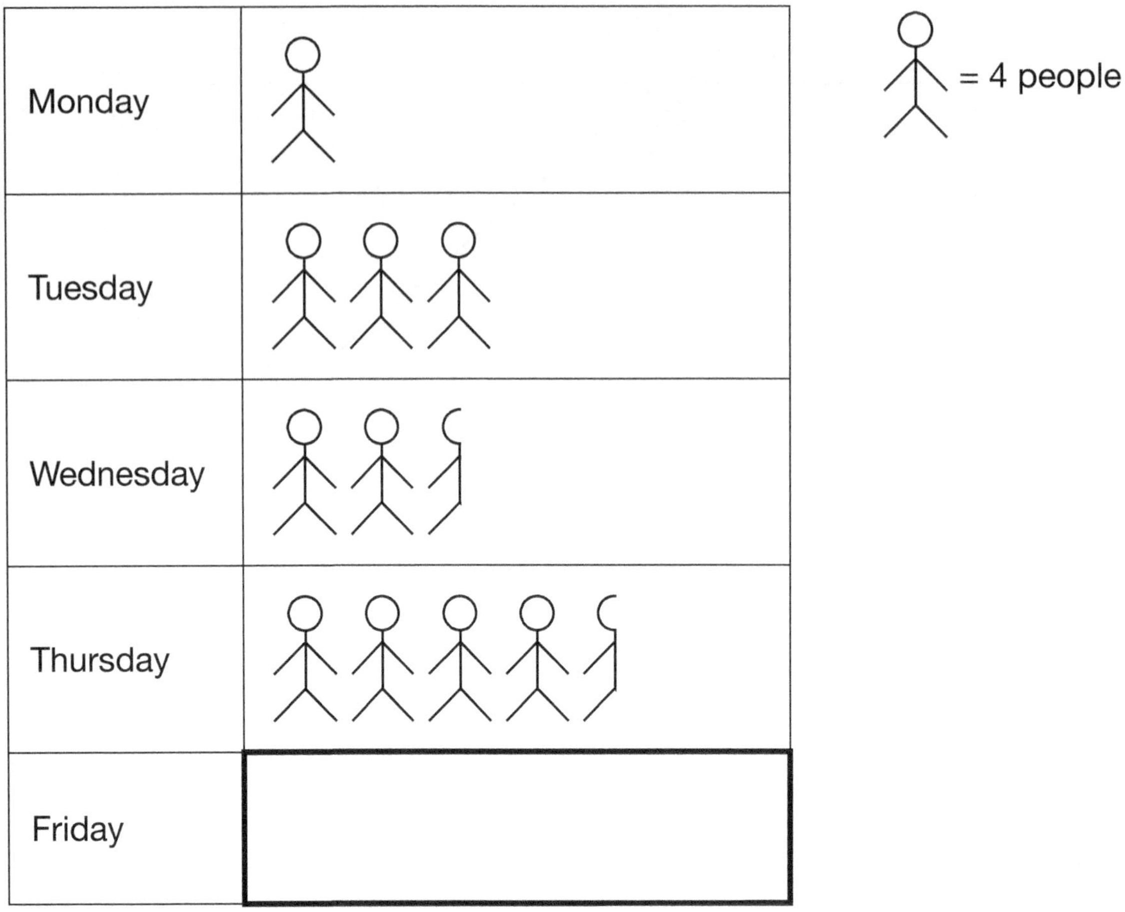

Monday	
Tuesday	
Wednesday	
Thursday	
Friday	

= 4 people

How many children attended the club on Tuesday?

1 mark

14 children attended the club on Friday.

Show this on the pictogram.

1 mark

14 Shade 2 more squares to create a symmetrical pattern.

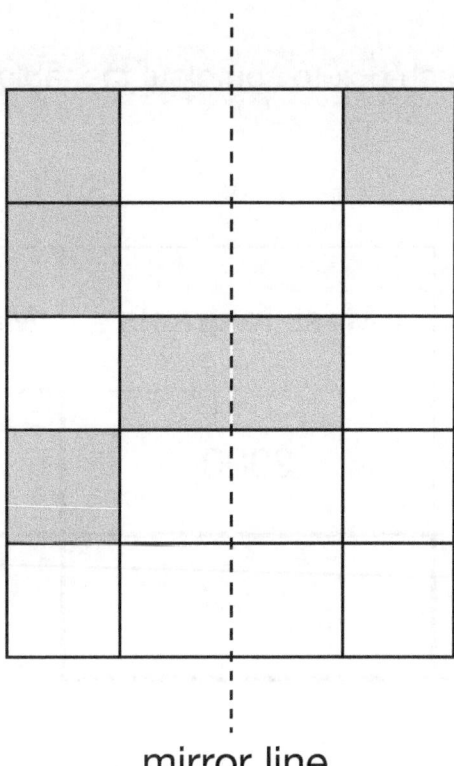

mirror line

1 mark

15 Draw an arrow (↓) to show the position of 0·23 on the number line.

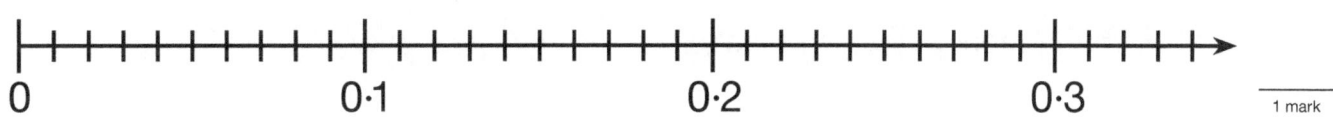

0   0·1   0·2   0·3

1 mark

Write the number that is $\frac{1}{100}$ more than 0·23

1 mark

16 The table shows the mass of some fruits.

Write **one number** in each box to complete the table.

	Mass in grams	Mass in kilograms
a bag of cherries	2000	
2 pineapples		1·6

2 marks

17 Laura has 12 cakes to sell for charity.

She sells 10 cakes for £4.80 each.

The remaining cakes are sold at half the price.

How much money does she make?

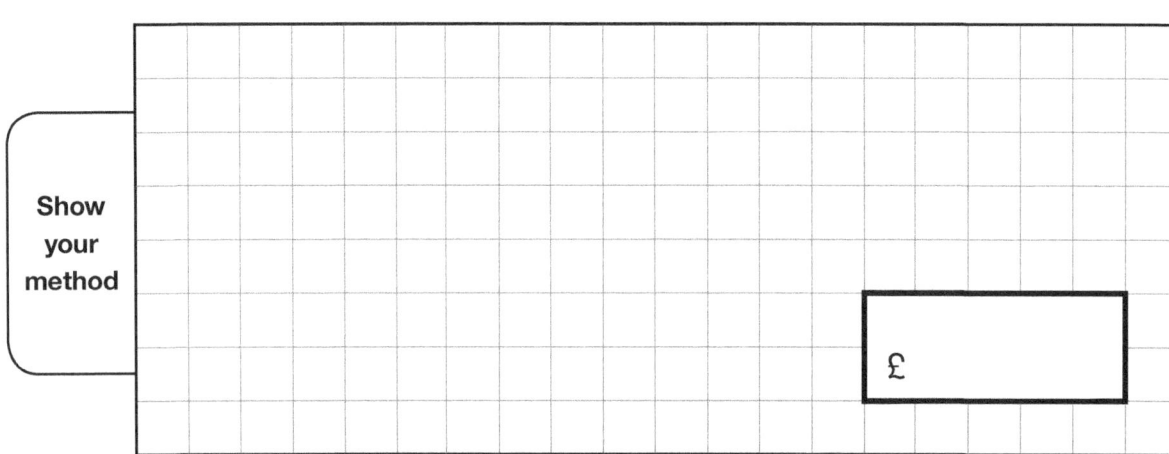

Show
your
method

£

2 marks

18 A square is drawn on the grid.

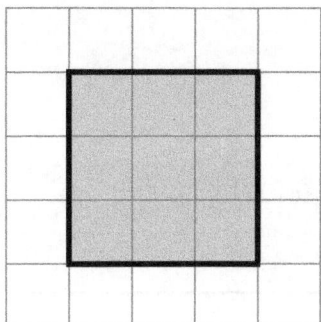

Tick (✔) the shapes with the **same area** as the square.

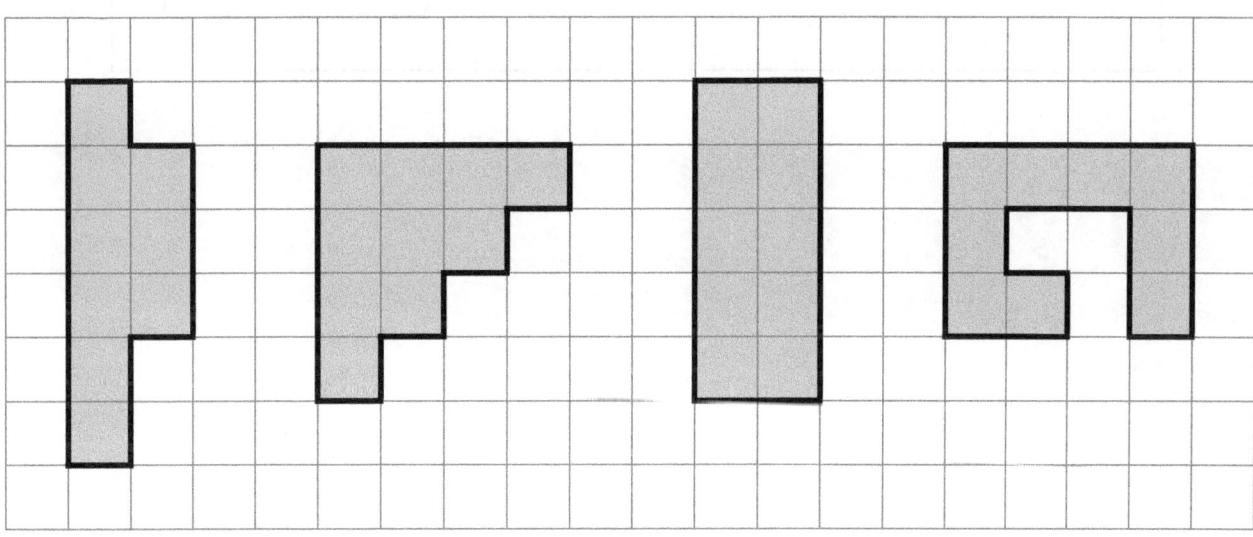

1 mark

19 Mike has 86 oranges.

He can pack 7 oranges into each box.

The last box is not full.

How many oranges are in the last box?

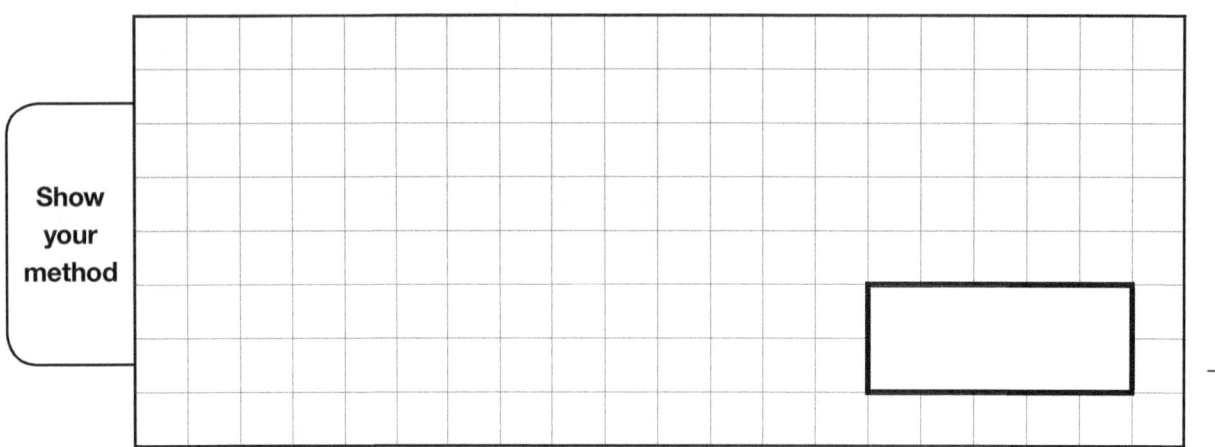

Show your method

2 marks

20 Thomas used his shoes to measure the lengths of some different cars.

Thomas's shoes are 9 cm long.

Tick (✔) the cars that have lengths between 150 cm and 200 cm

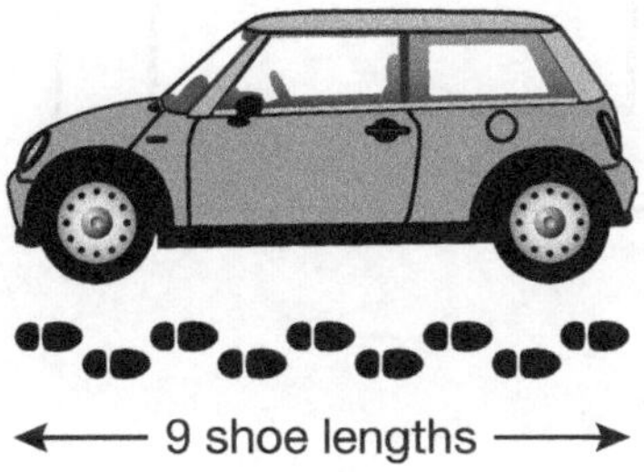

⟵ 9 shoe lengths ⟶

☐

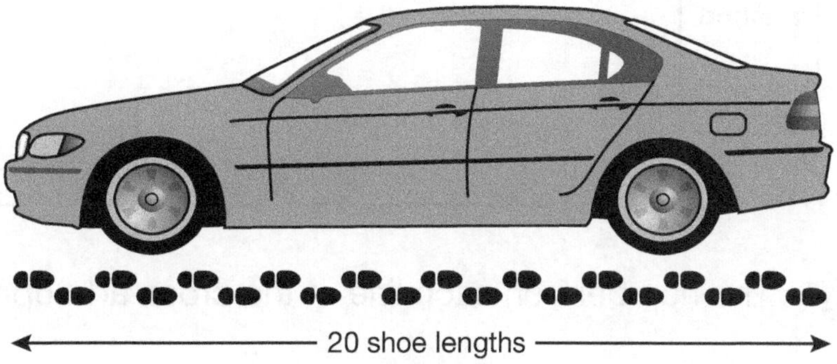

⟵ 20 shoe lengths ⟶

☐

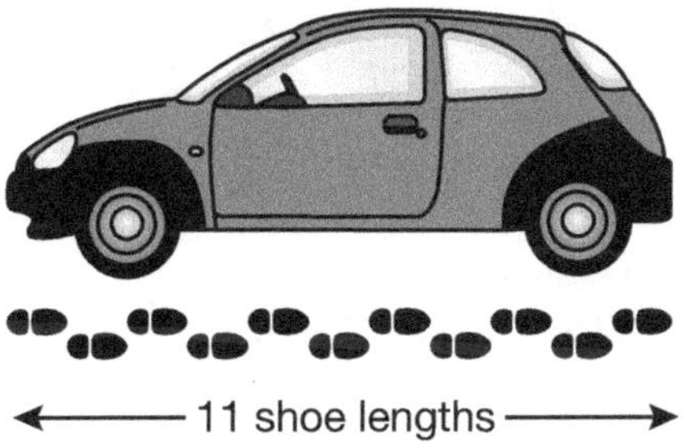

⟵ 11 shoe lengths ⟶

☐

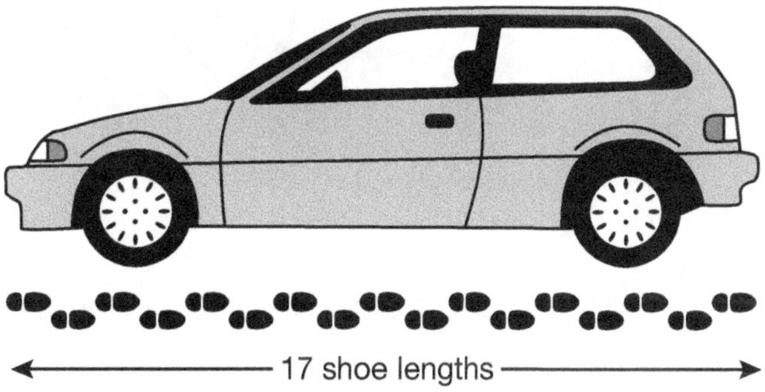

⟵ 17 shoe lengths ⟶

☐

1 mark

21 Part of the 37 multiplication table is shown below.

$$1 \times 37 = 37$$
$$2 \times 37 = 74$$
$$3 \times 37 = 111$$
$$4 \times 37 = 148$$

Use the **table** to find the answer to 8×37

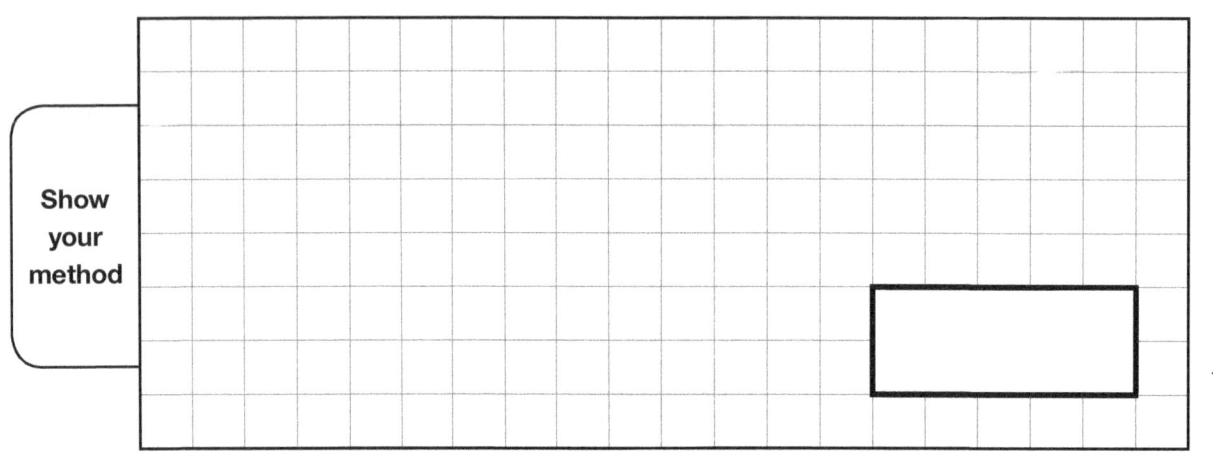

Show your method

2 marks

22 The numbers on each line of the cross add up to 140

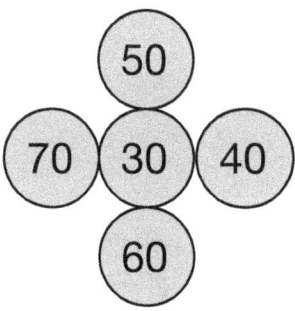

50

70 30 40

60

Rearrange the numbers so that the numbers on each line of the cross add up to 160

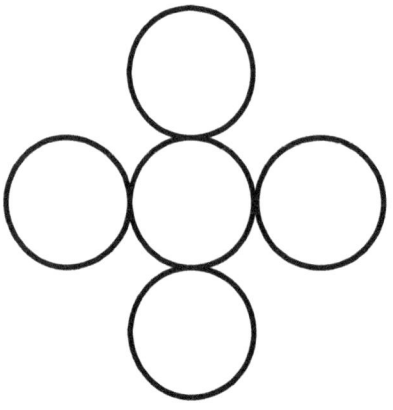

1 mark

Reasoning

Name: _____

Class: _____

Instructions

You **may not** use a calculator to answer any questions in this test.

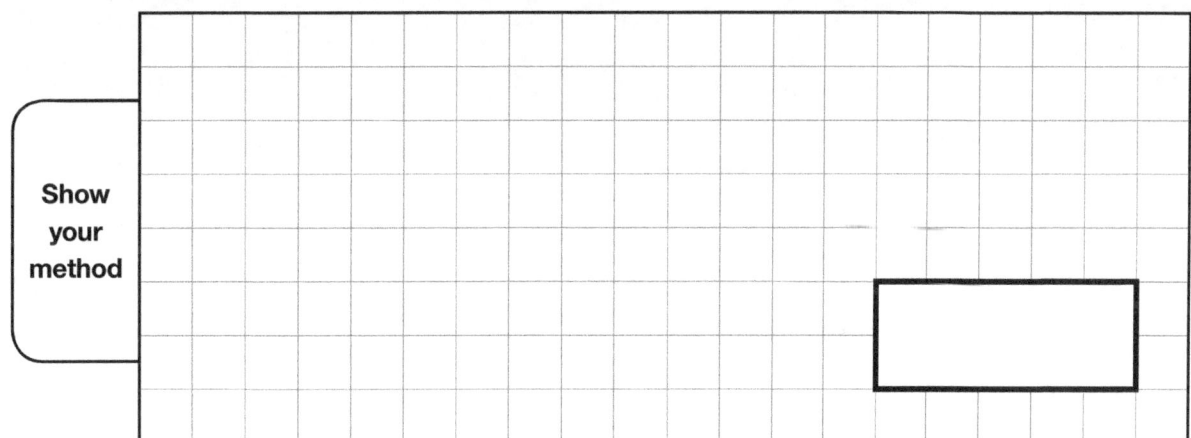

1 Look at the two cards.

3 thousands
7 hundreds
5 tens

6 thousands
11 hundreds
0 tens

Tick (✔) the statement that correctly compares these two numbers.

6110 < 3705 ☐   7100 > 3705 ☐

6110 < 3750 ☐   7100 > 3750 ☐

1 mark

2 Tick (✔) the flags with **exactly one** line of symmetry.

☐

☐

☐

☐

1 mark

3 | Write **one number** in each box to complete the sequence.

[   ]   4163   5163   [   ]   7163

1 mark

4   The cat is chasing mice.

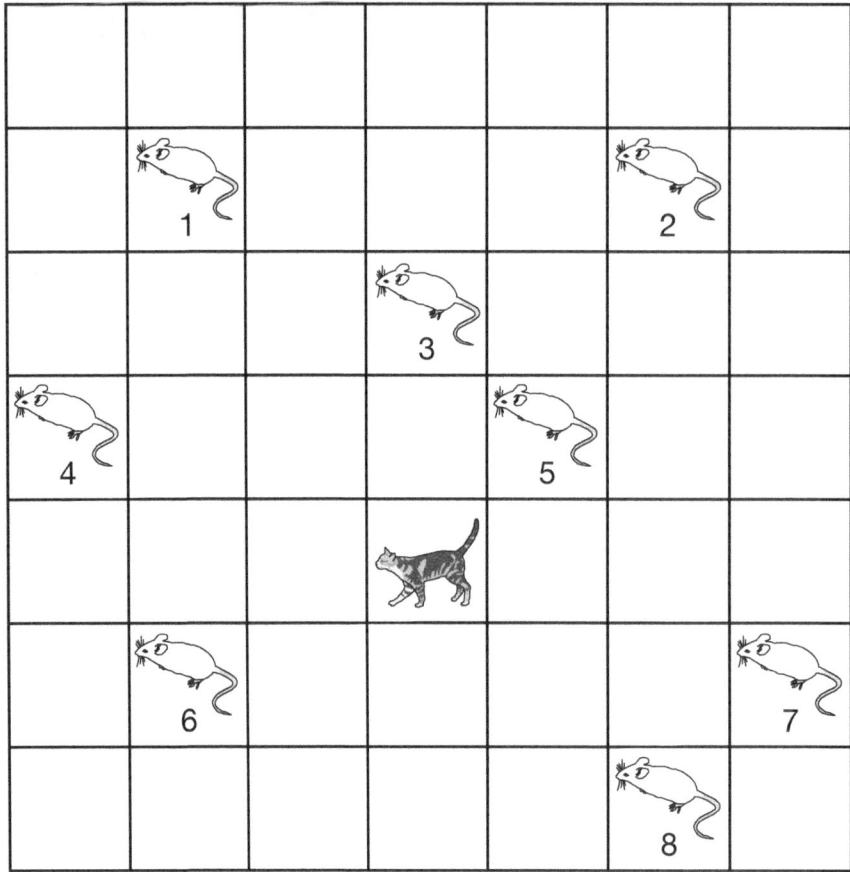

The cat moves 2 squares right and 3 squares up.

Write the number of the mouse that it catches.

1 mark

Write **one number** in each box to complete the calculations.

$$35 \div \boxed{} = 3{\cdot}5$$

$$\boxed{} \div 100 = 6{\cdot}5$$

1 mark

Write **all** the missing numbers in this multiplication grid.

×	3		9
2	6		18
		49	
	33	77	99

2 marks

7 Draw an arrow (↓) to show the Roman numeral XL on the number line.

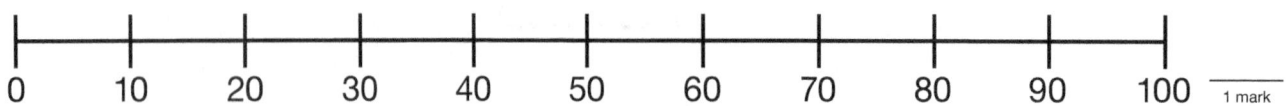

1 mark

8 Write the coordinates of the point marked with a cross (✗).

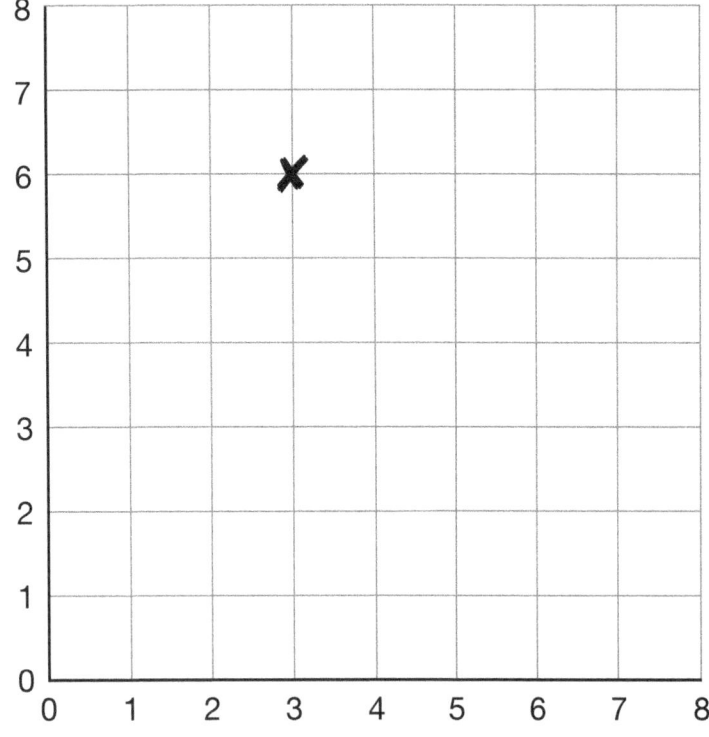

(,   )

1 mark

9 There are 143 students in a school.

The Art department has £800 to spend.

They spend £5 per student on materials.

How much do they have left?

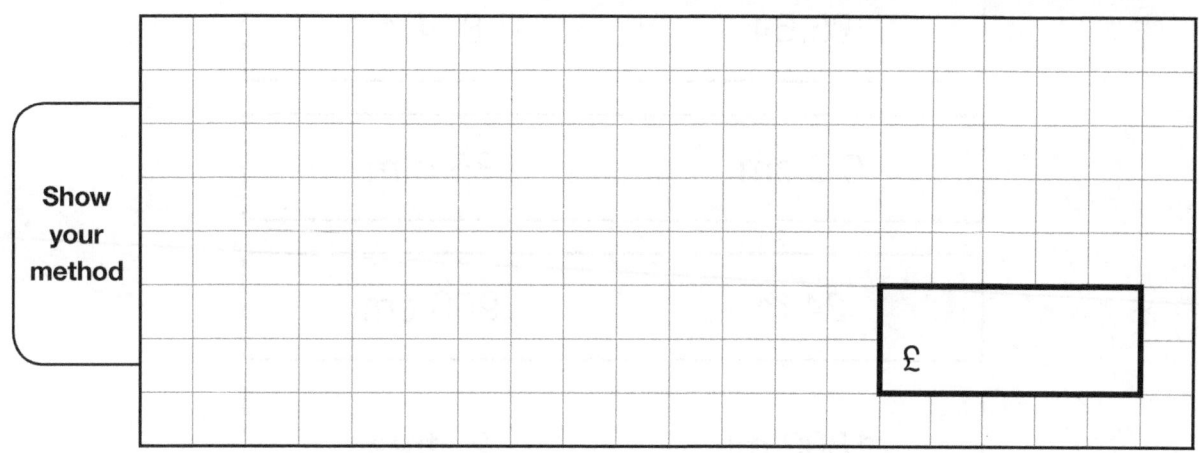

Show your method

£

2 marks

10 Write a number in each box to make a correct calculation.

$$\boxed{} \times \boxed{} \times \boxed{} = 100$$

1 mark

11 Circle the **larger** measure in each pair.

One has been done for you.

(650 g)	300 g

£5.64	814 p

6·3 cm	85 mm

24 m	960 cm

1800 ml	2 litres

2 marks

12 The graph shows the temperature in °C at different times in one day.

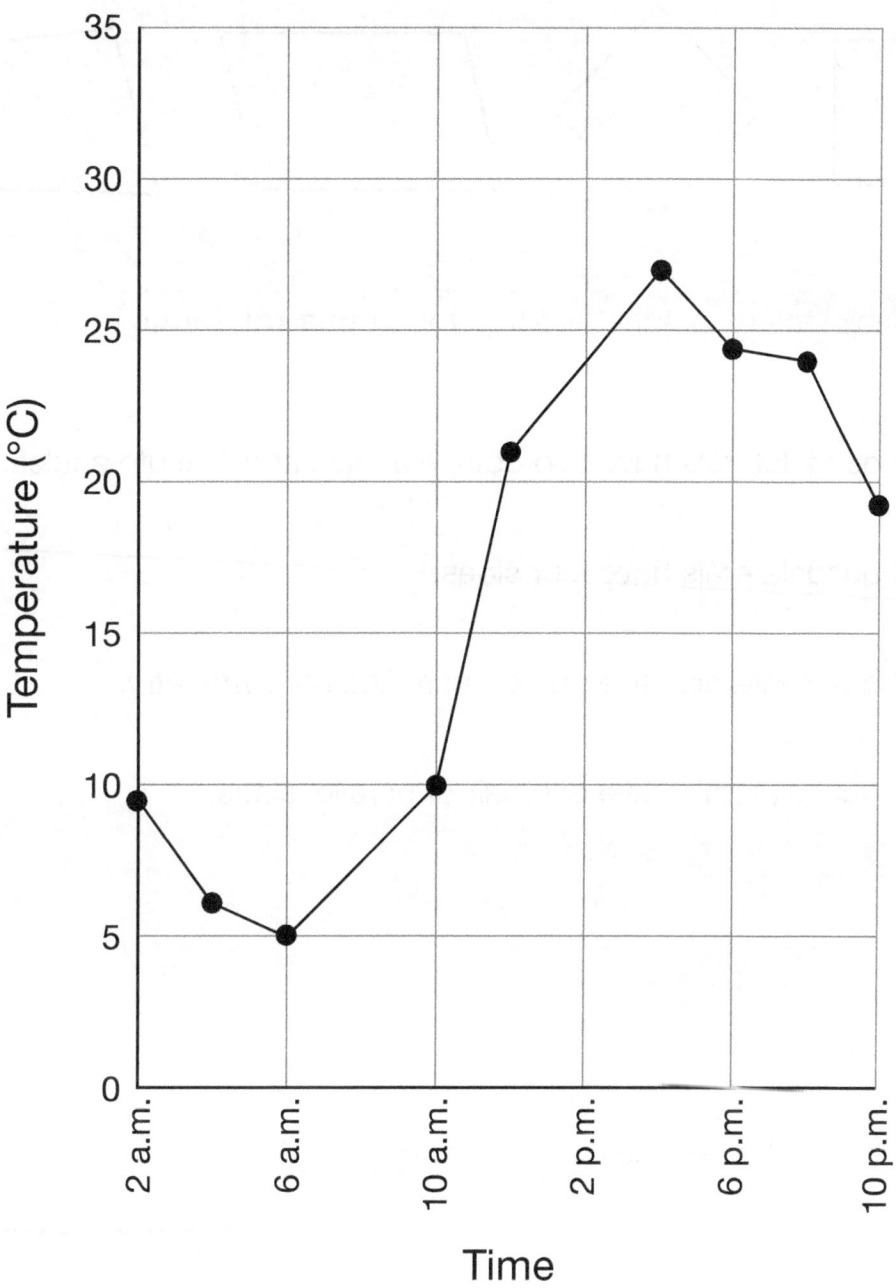

What was the temperature at 10 a.m.?

°C

Between which hours of the day was the temperature increasing?

13 Some quadrilaterals are shown below.

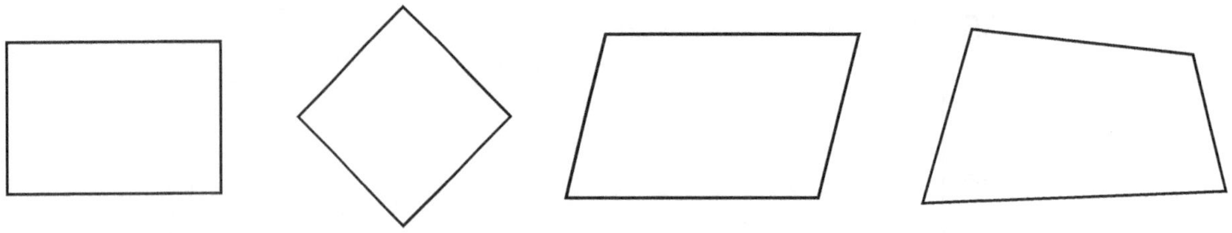

Tick (✔) any statements that are true for **all quadrilaterals**.

☐ All quadrilaterals have two obtuse angles and 2 acute angles.

☐ All quadrilaterals have four sides.

☐ All quadrilaterals have one or more lines of symmetry.

☐ All quadrilaterals have one pair of parallel sides.

1 mark

14 Write a number that is between 6·46 and 6·64

1 mark

15 A trapezium is made from three equilateral triangles.

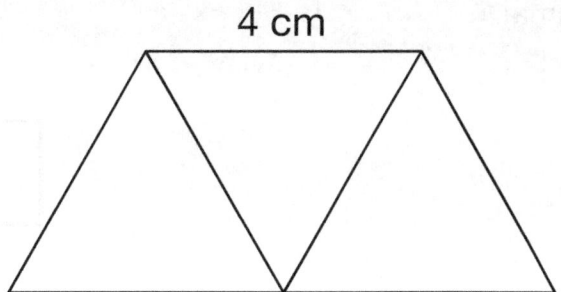

4 cm

What is the perimeter of the trapezium?

| cm |

16 Circle the number that is a multiple of 4 **and** a multiple of 6

40	41	42	43
44	45	46	47
48	49	50	51
52	53	54	55

17 Jonas measures the length of a classroom as $4\frac{3}{10}$ metres.

Write $4\frac{3}{10}$ as a decimal.

1 mark

18 There are 7 people in a netball team.

105 children sign up for a netball competition.

How many teams can be made?

1 mark

19 $\frac{1}{4}$ of the shape is shaded.

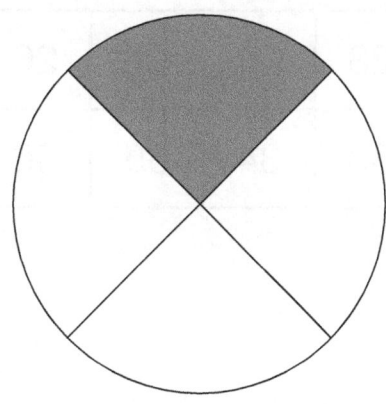

Write two fractions equivalent to $\frac{1}{4}$

[ ] and [ ]

2 marks

20 Draw a rectangle with an area of 15 square centimetres.

1 mark

21 Shade **all** the numbers that are 30 when rounded to the nearest 10

20	21	22	23	24	25	26	27	28	29
30	31	32	33	34	35	36	37	38	39

1 mark

22 Two numbers have a difference of 185

One of the numbers is 4560

Find the other **two** possible numbers.

[   ] and [   ]

2 marks

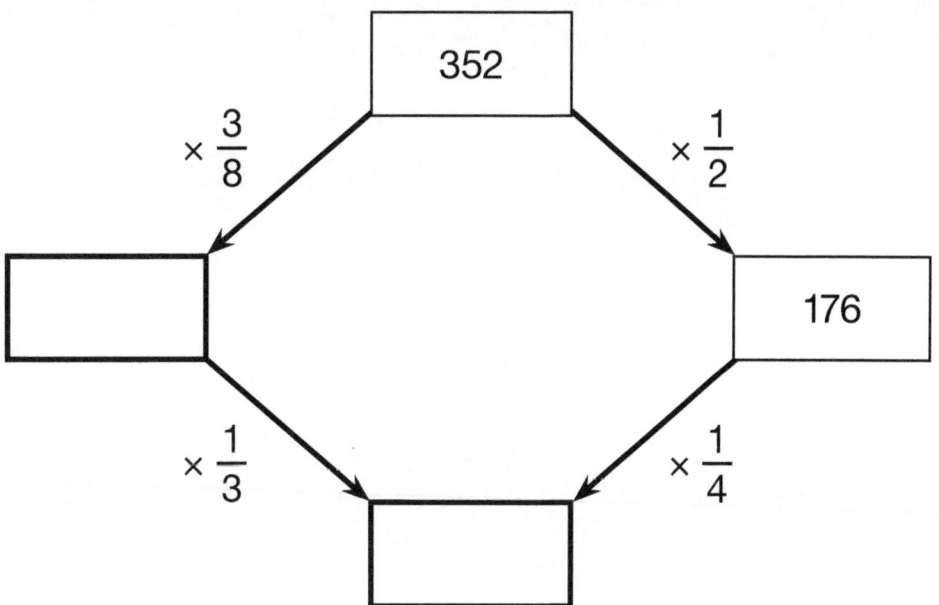

2 marks

Arithmetic

Instructions

You **may not** use a calculator to answer any questions in this test.

Questions and answers

You have **25 minutes** to complete this test.

Work as quickly and as carefully as you can.

Put your answer in the box for each question.

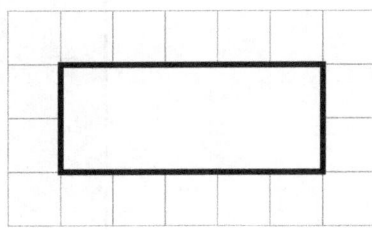

For questions expressed as common fractions, you should give your answer as common fractions.

All other answers should be given as whole numbers or decimals.

If you cannot do one of the questions, **go on to the next one**. You can come back to it later if you have time.

If you finish before the end, **go back and check your work**.

Marks

The number under each box at the side of the page tells you the maximum number of marks for each question.

In this test, some multiplication questions are worth **2 marks each**. You will be awarded 2 marks for a correct answer. You may get 1 mark for showing the formal method.

All other questions are worth **1 mark each**.

1

$8 \times 6 =$

48

1 mark

2

$3{\cdot}6 + 4{\cdot}1 =$

7·7

1 mark

3

$78 + 1000 =$

1078

1 mark

4

$$\frac{9}{11} - \frac{7}{11} =$$

$$\frac{2}{11}$$

1 mark

5

2147 + 52 =

2199

1 mark

6

30 ÷ 6 =

5

1 mark

7   125 + 25 + 25 + 25 =

	200

1 mark

8 $\frac{1}{9}$ of 36 =

	4

1 mark

9   5604 − 800 =

	4804

1 mark

10 $132 \div 11 =$

12

1 mark

11 $\dfrac{2}{3} + \dfrac{2}{3} =$

$1\frac{1}{3}$

1 mark

12 $32{\cdot}6 + 28{\cdot}7 =$

61·3

1 mark

13 $7 \times 3 \times 4 =$

84

14 $6{\cdot}7 + 1{\cdot}03 =$

7·73

15 $756 + 364 =$

1120

16 43 × 5 =

215

1 mark

17 8 ÷ 10 =

0·8

1 mark

18 4173 − 992 =

3181

1 mark

19 $1 - \dfrac{1}{6} =$

$\dfrac{5}{6}$

1 mark

20 $330 \div 3 =$

110

1 mark

21 £1.60 + £2.45 =

£4.05

1 mark

22						
		2	6			
	×		8			

Show your method

208

2 marks

23 1000 − 79 =

921

1 mark

24 3106 + 2714 =

5820

1 mark

25

$\frac{5}{6}$ of 72 =

60

1 mark

26

Show your method

$$\begin{array}{r} 3\ 1\ 7 \\ \times\quad\quad 4 \\ \hline \end{array}$$

1268

2 marks

27

7 ÷ 100 =

0·07

1 mark

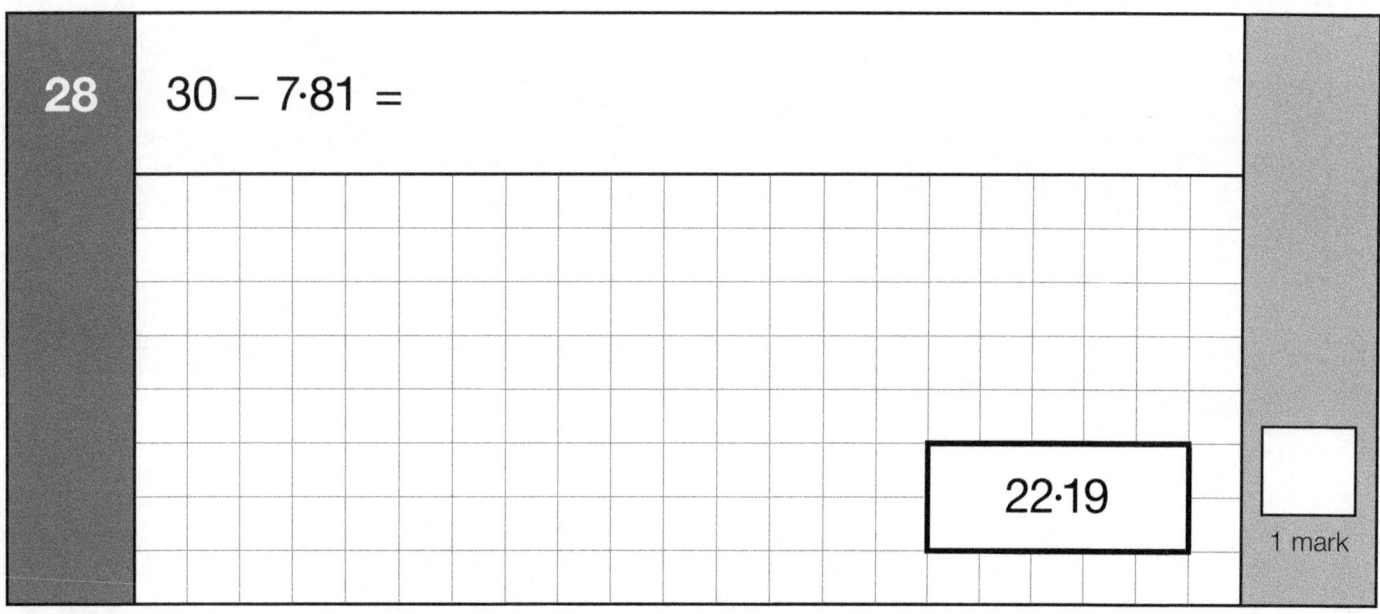

28

$30 - 7 \cdot 81 =$

22·19

1 mark

Reasoning

Instructions

You **may not** use a calculator to answer any questions in this test.

Questions and answers

You have **30 minutes** to complete this test.

Follow the instructions for each question.

Work as quickly and as carefully as you can.

If you need to do working out, you can use the space around the question.

Some questions have a method box like this:

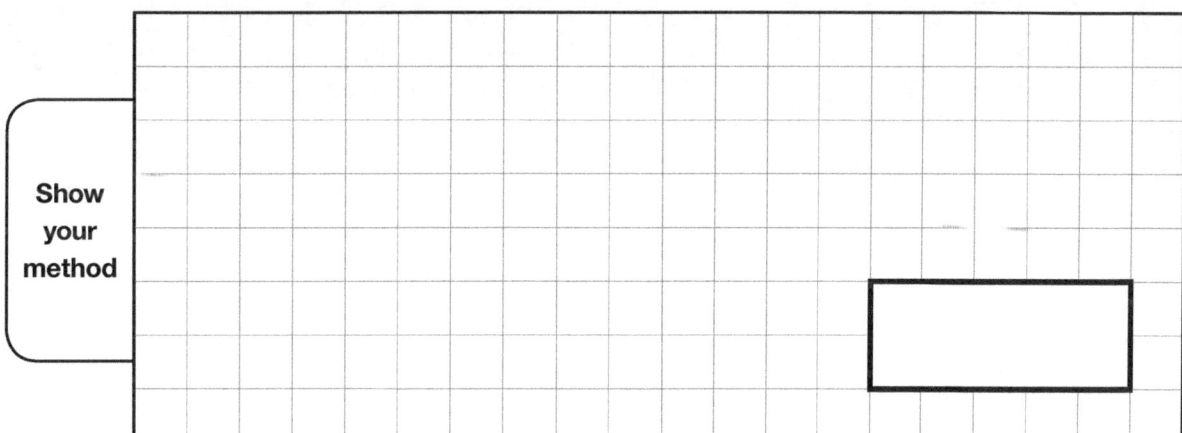

Show
your
method

For these questions you may get a mark for showing your method.

If you cannot do one of the questions, **go to the next one**.
You can come back to it later, if you have time.

If you finish before the end, **go back and check your work**.

Marks

The number under each line at the side of the page tells you the maximum number
of marks for each question.

1 Circle the **largest** number in each pair.

One has been done for you.

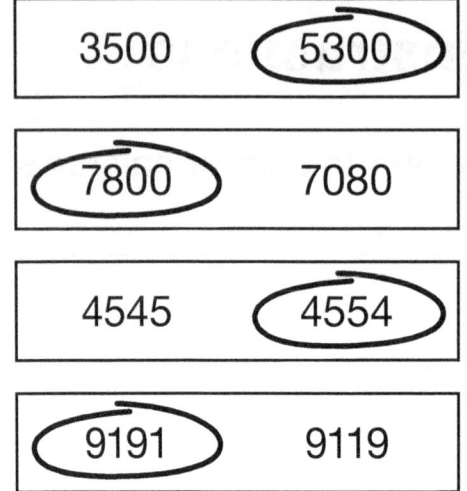

1 mark

Draw lines to join each pair of numbers with a total of 1000

One has been done for you.

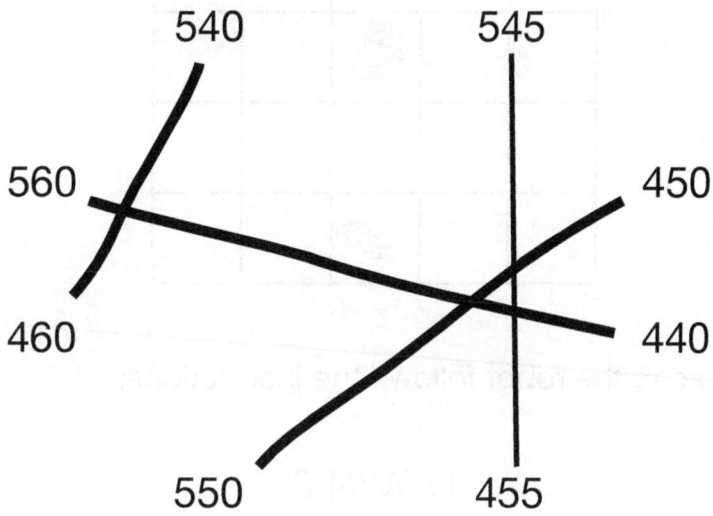

540   545

560   450

460   440

550   455

1 mark

3 A robot is in the middle of the grid.

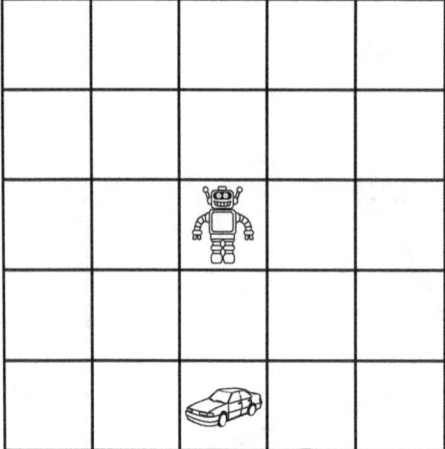

To move to the car, the robot follows the instructions:

DOWN 2

The robot is in a new position on the grid.

Write instructions to move the robot to the **tree**.

Use only **UP** or **DOWN**, and **LEFT** or **RIGHT**.

RIGHT 1   UP 3

4 The temperature is 7 degrees Celsius.

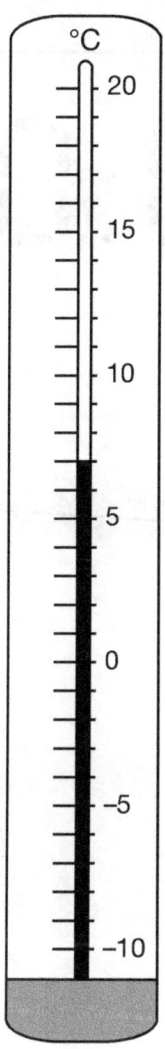

The temperature falls by 11 degrees Celsius.

What is the new temperature?

–4 °C

5 This is the time on a digital clock.

Tick (✔) the same time on an analogue clock.

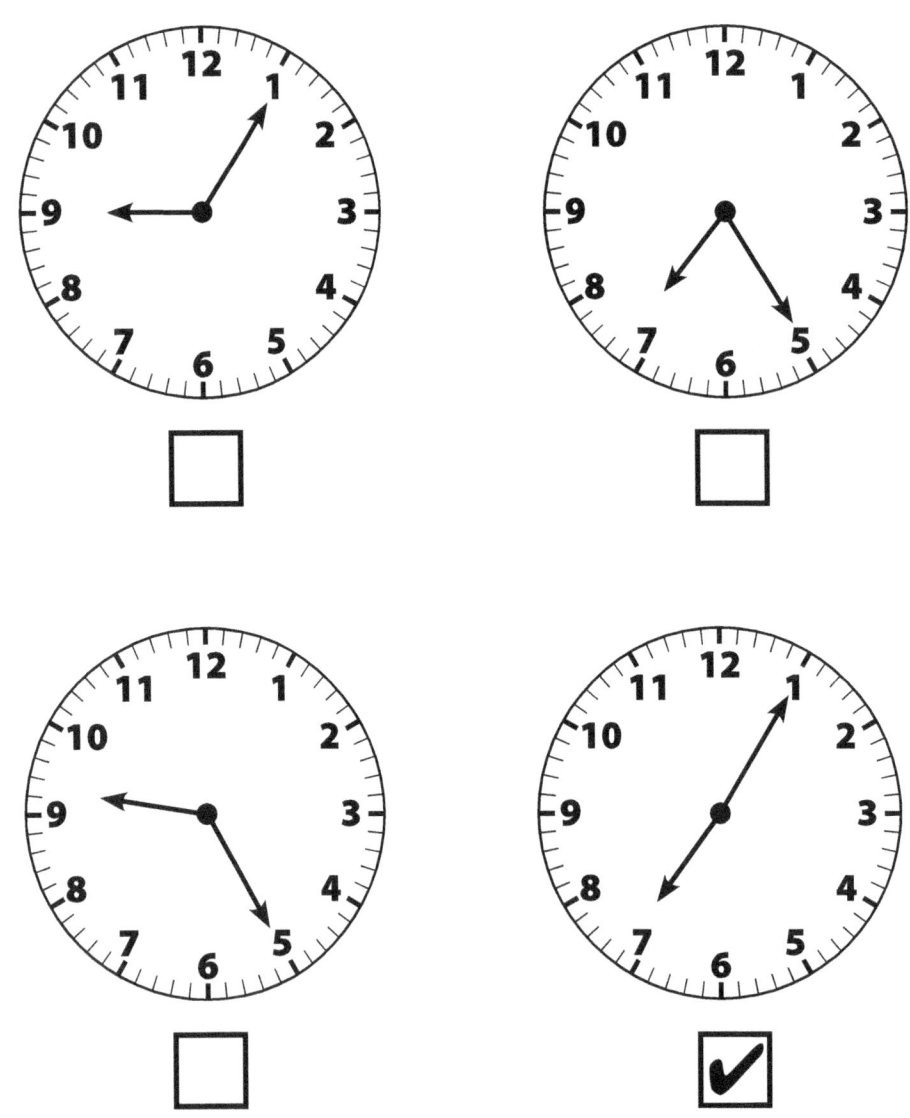

1 mark

6 Tick (✔) the shape that can be cut along one of its lines of symmetry to create two trapeziums.

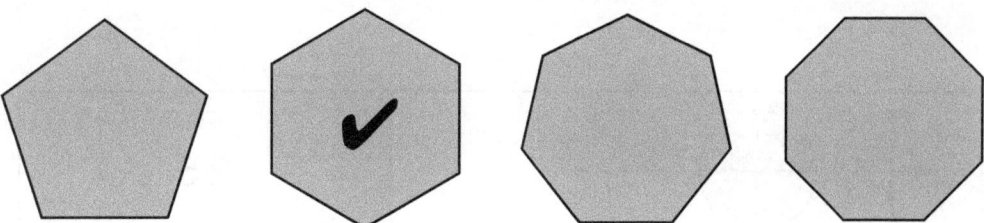

1 mark

7 Circle all the numbers that are **not** in the 7 multiplication table.

 49   56   70   84

1 mark

Write the decimal equivalents shown by these shapes.

One has been done for you.

	2·5
	3·25
	1·75

2 marks

9 A stadium can hold 8000 people when it is full.

The table below shows the number of people that attended a concert one weekend.

Day	Number of people
Saturday	6392
Sunday	7239

How many **more** people could have attended altogether?

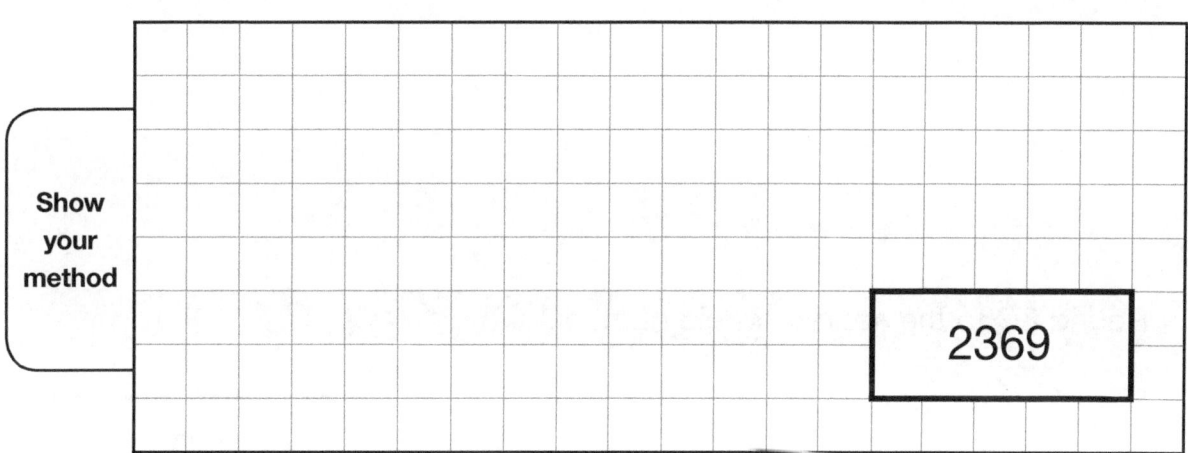

Show your method

2369

2 marks

10 Complete the calculation.

$$3\frac{4}{5} + \boxed{6\frac{1}{5}} = 10$$

1 mark

11 Round 8·7 to the nearest whole number.

$$\boxed{9}$$

1 mark

12 Four angles are marked on the diagram.

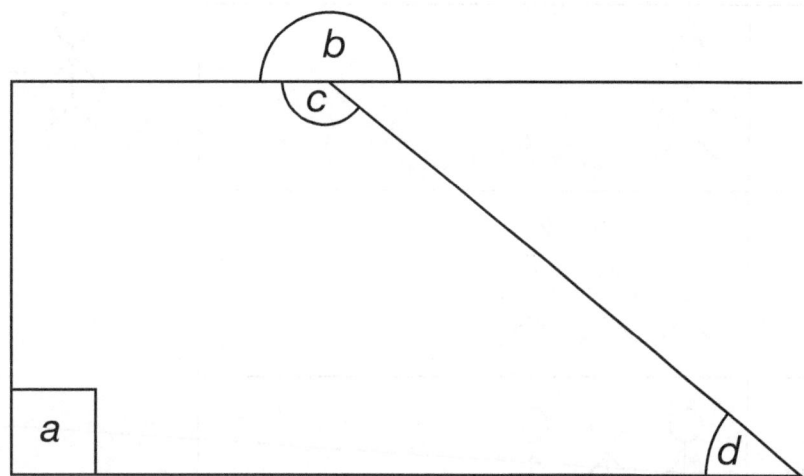

Match the letter of each angle to its name.

One has been done for you.

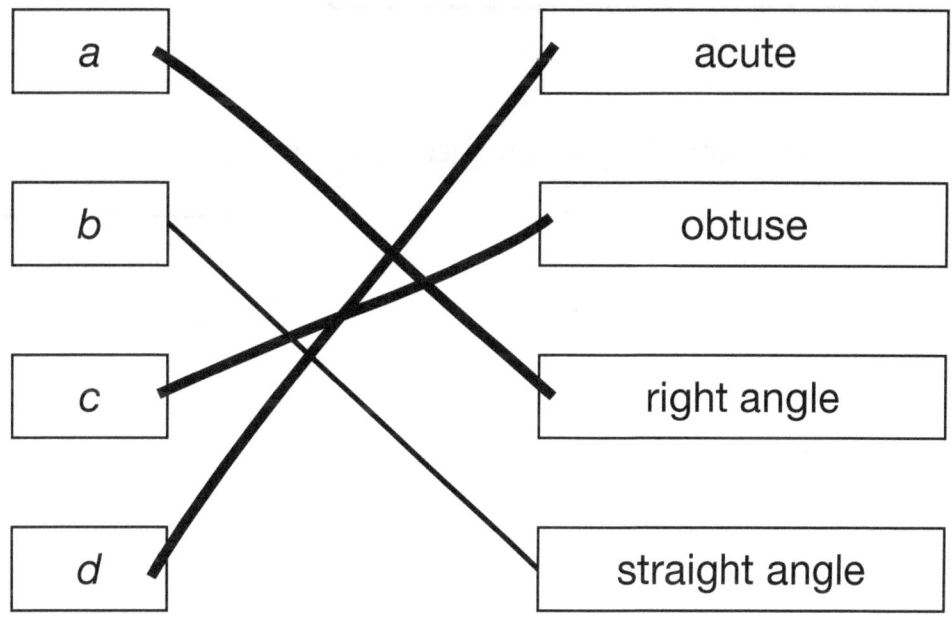

1 mark

13 A teacher counted the number of children attending a club each day for a week.

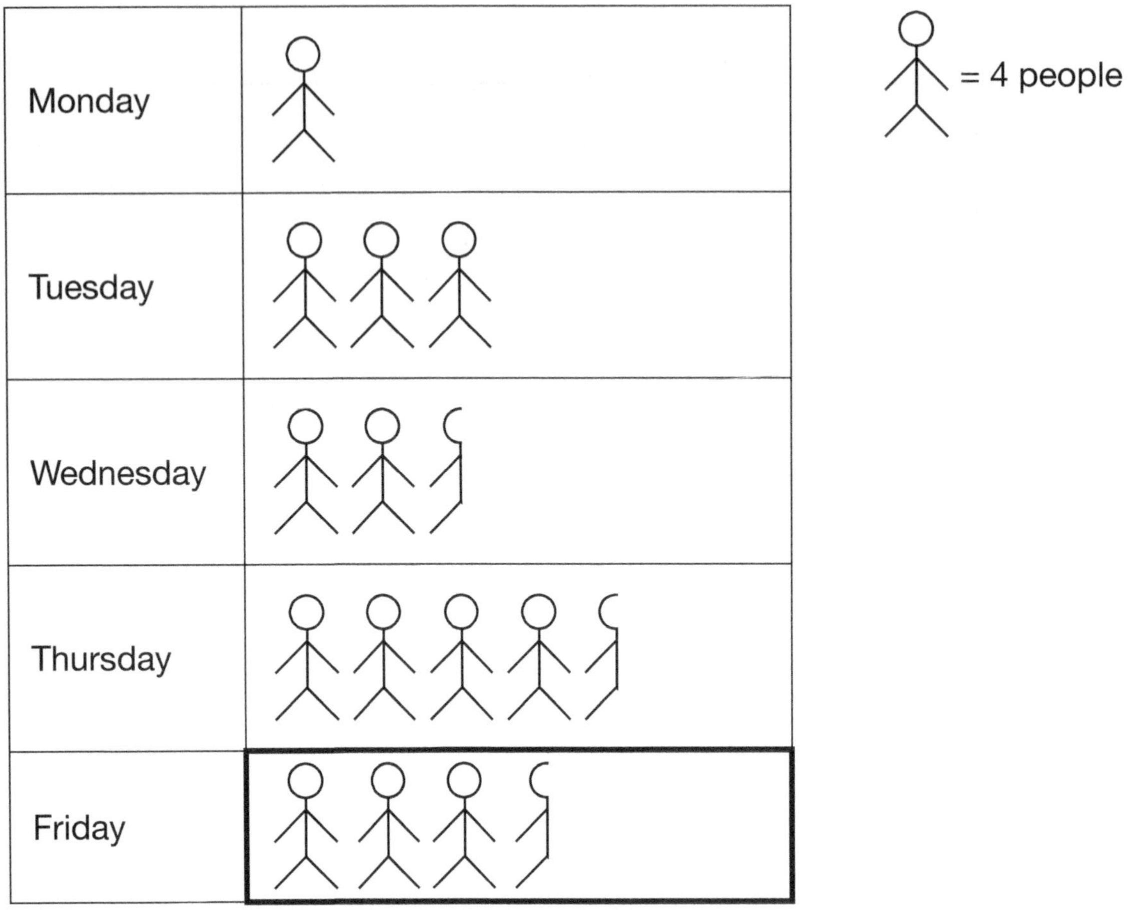

Monday	
Tuesday	
Wednesday	
Thursday	
Friday	

= 4 people

How many children attended the club on Tuesday?

12

1 mark

14 children attended the club on Friday.

Show this on the pictogram.

1 mark

14 Shade 2 more squares to create a symmetrical pattern.

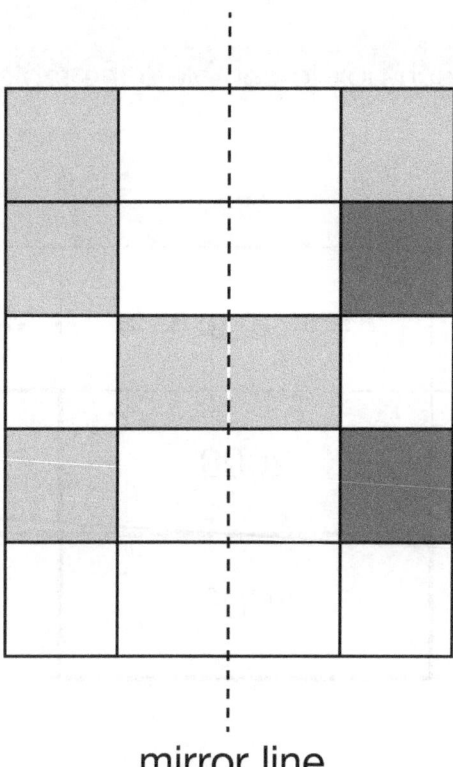

mirror line

1 mark

15 Draw an arrow (↓) to show the position of 0·23 on the number line.

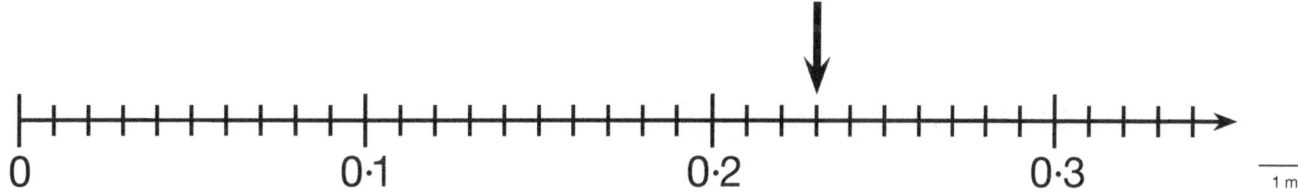

0   0·1   0·2   0·3

1 mark

Write the number that is $\frac{1}{100}$ more than 0·23

0·24

1 mark

16 The table shows the mass of some fruits.

Write **one number** in each box to complete the table.

	Mass in grams	Mass in kilograms
a bag of cherries	2000	2
2 pineapples	1600	1·6

2 marks

17 Laura has 12 cakes to sell for charity.

She sells 10 cakes for £4.80 each.

The remaining cakes are sold at half the price.

How much money does she make?

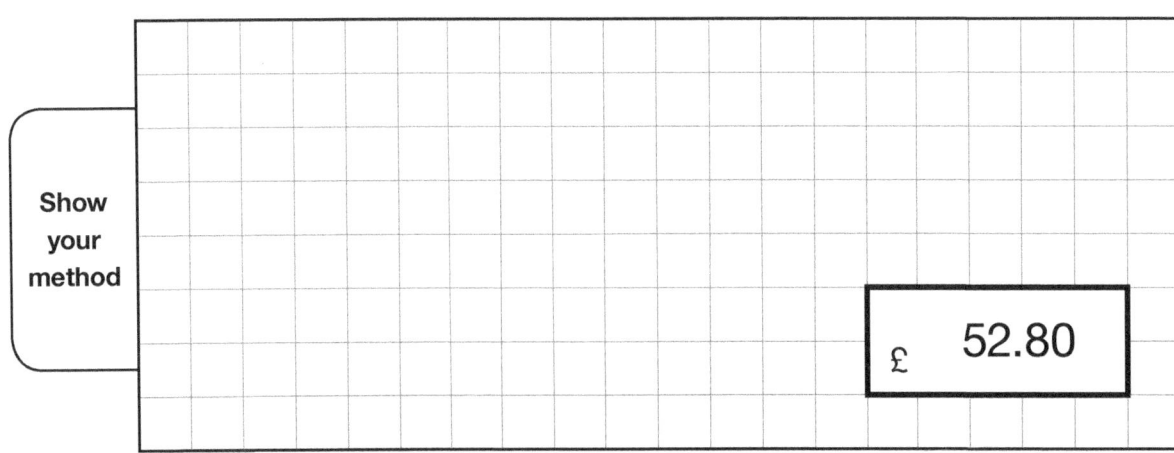

Show your method

£ 52.80

2 marks

18   A square is drawn on the grid.

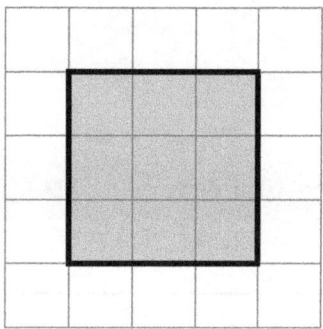

Tick (✔) the shapes with the **same area** as the square.

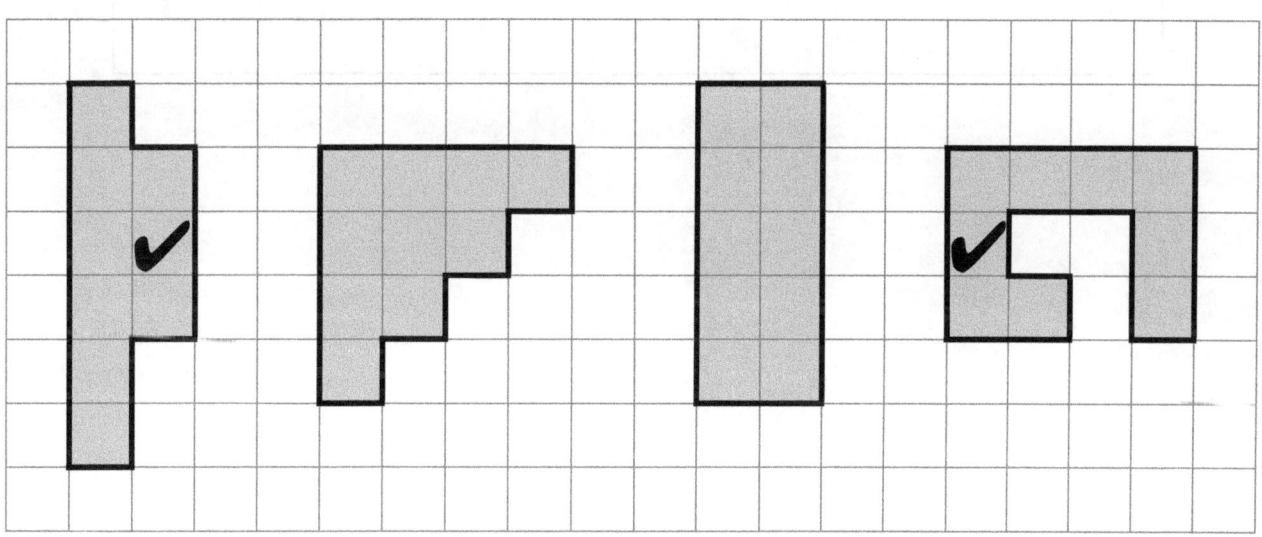

1 mark

19 Mike has 86 oranges.

He can pack 7 oranges into each box.

The last box is not full.

How many oranges are in the last box?

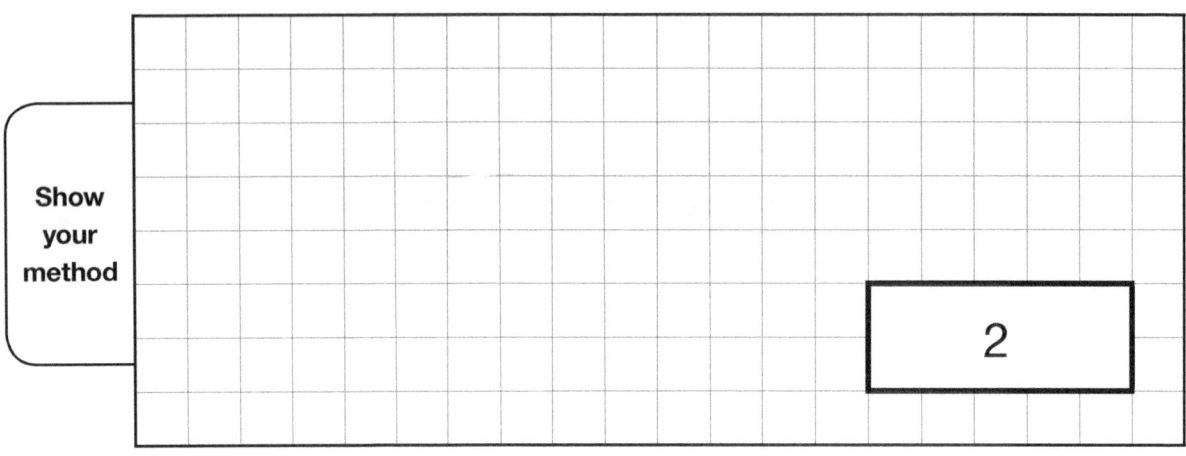

Show
your
method

2

2 marks

20 Thomas used his shoes to measure the lengths of some different cars.

Thomas's shoes are 9 cm long.

Tick (✔) the cars that have lengths between 150 cm and 200 cm

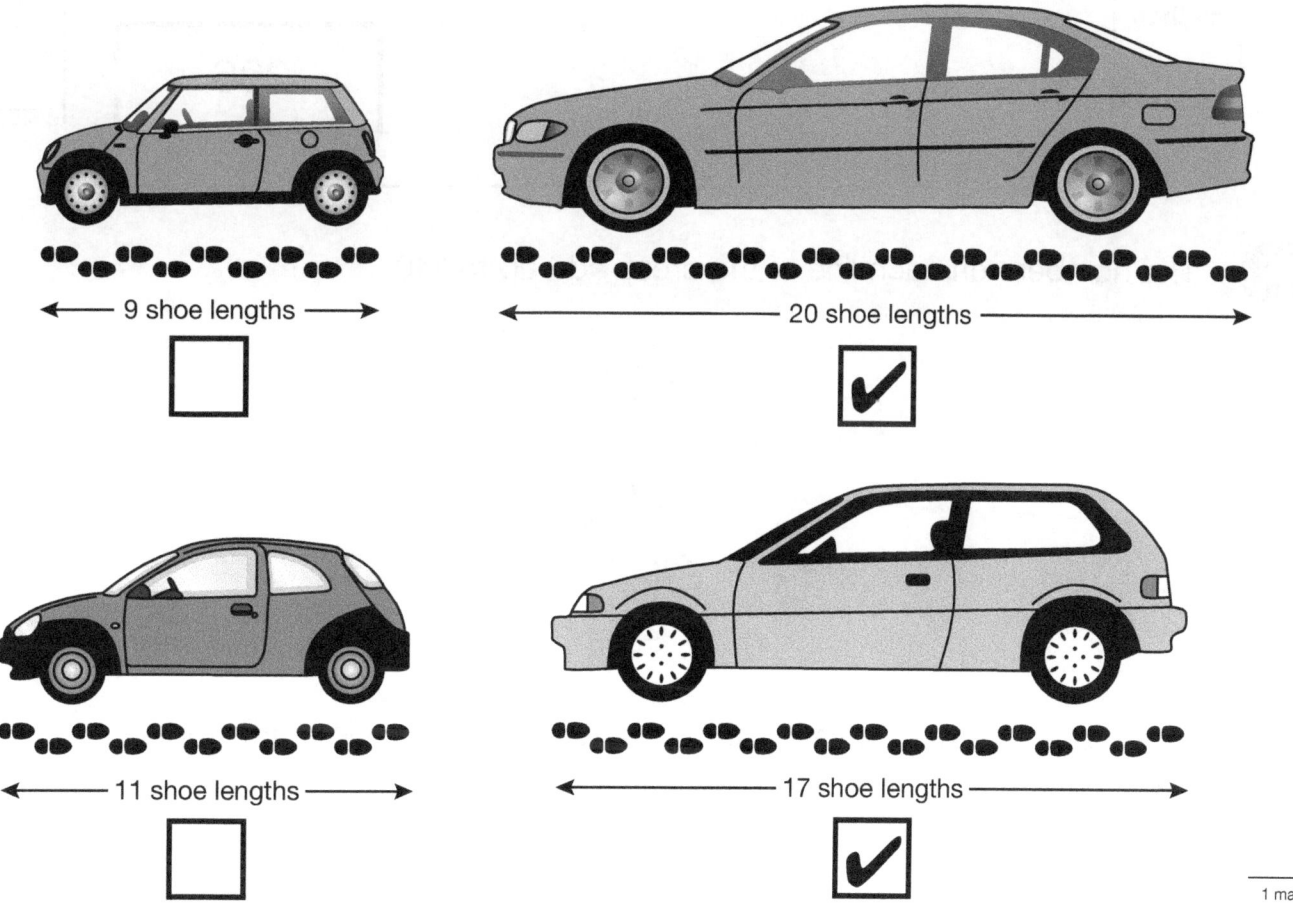

←—— 9 shoe lengths ——→

[ ]

←————— 20 shoe lengths —————→

[✔]

←——— 11 shoe lengths ———→

[ ]

←————— 17 shoe lengths —————→

[✔]

1 mark

21 Part of the 37 multiplication table is shown below.

$$1 \times 37 = 37$$
$$2 \times 37 = 74$$
$$3 \times 37 = 111$$
$$4 \times 37 = 148$$

Use the **table** to find the answer to 8 × 37

Show
your
method

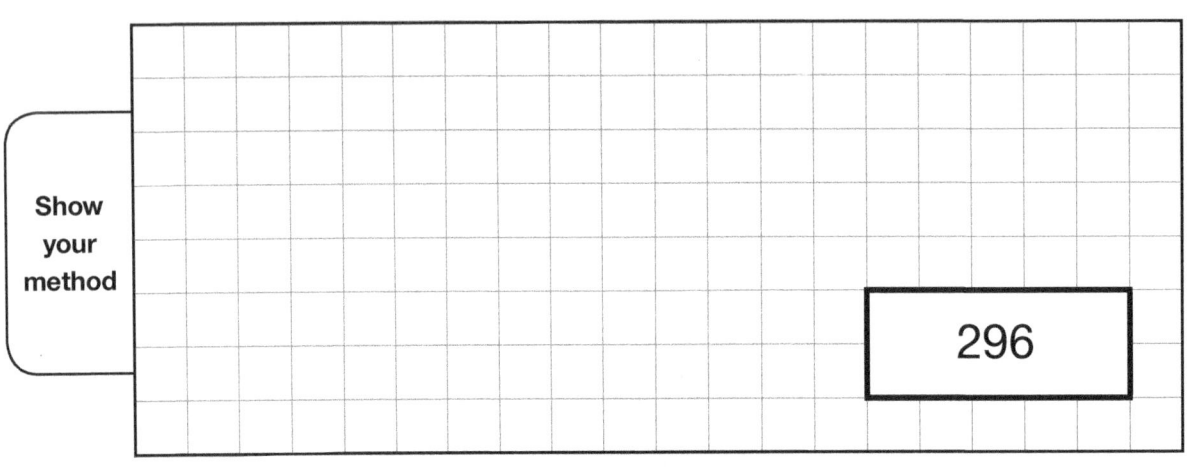

296

2 marks

22 The numbers on each line of the cross add up to 140

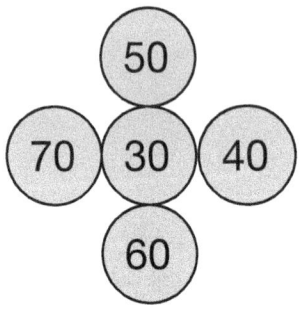

Rearrange the numbers so that the numbers on each line of the cross add up to 160

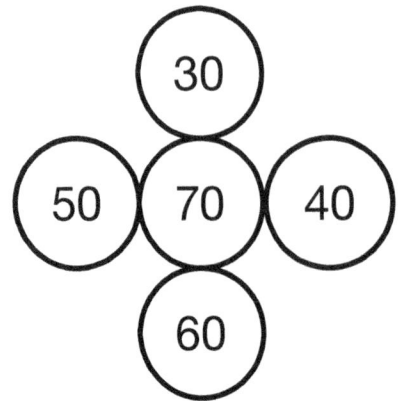

1 mark

Reasoning

Instructions

You **may not** use a calculator to answer any questions in this test.

Questions and answers

You have **30 minutes** to complete this test.

Follow the instructions for each question.

Work as quickly and as carefully as you can.

If you need to do working out, you can use the space around the question.

Some questions have a method box like this:

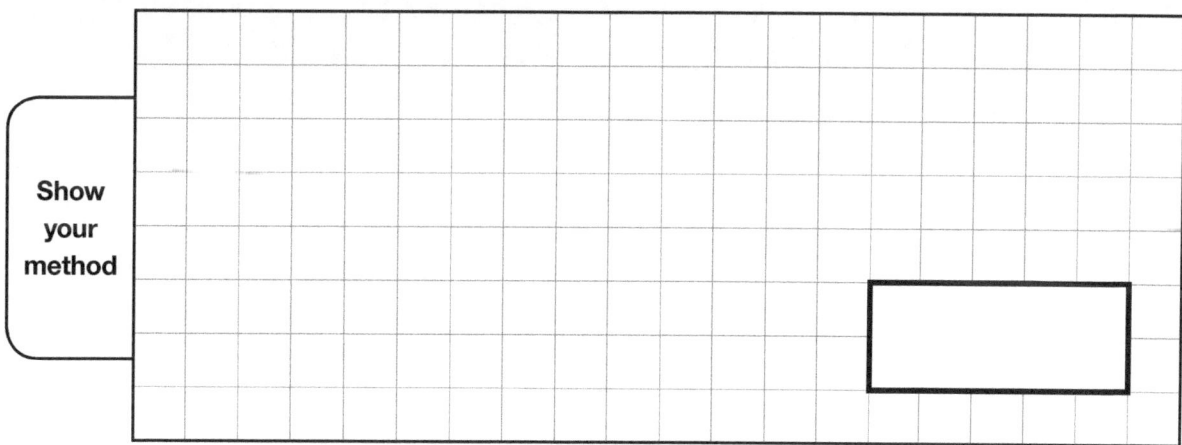

For these questions you may get a mark for showing your method.

If you cannot do one of the questions, **go to the next one.**
You can come back to it later, if you have time.

If you finish before the end, **go back and check your work.**

Marks

The number under each line at the side of the page tells you the maximum number of marks for each question.

1   Look at the two cards.

3 thousands
7 hundreds
5 tens

6 thousands
11 hundreds
0 tens

Tick (✔) the statement that correctly compares these two numbers.

6110 < 3705 ☐   7100 > 3705 ☐

6110 < 3750 ☐   7100 > 3750

1 mark

2 Tick (✔) the flags with **exactly one** line of symmetry.

☐

✔

✔

☐

1 mark

3 Write **one number** in each box to complete the sequence.

| 3163 | 4163 | 5163 | 6163 | 7163 |

1 mark

4 The cat is chasing mice.

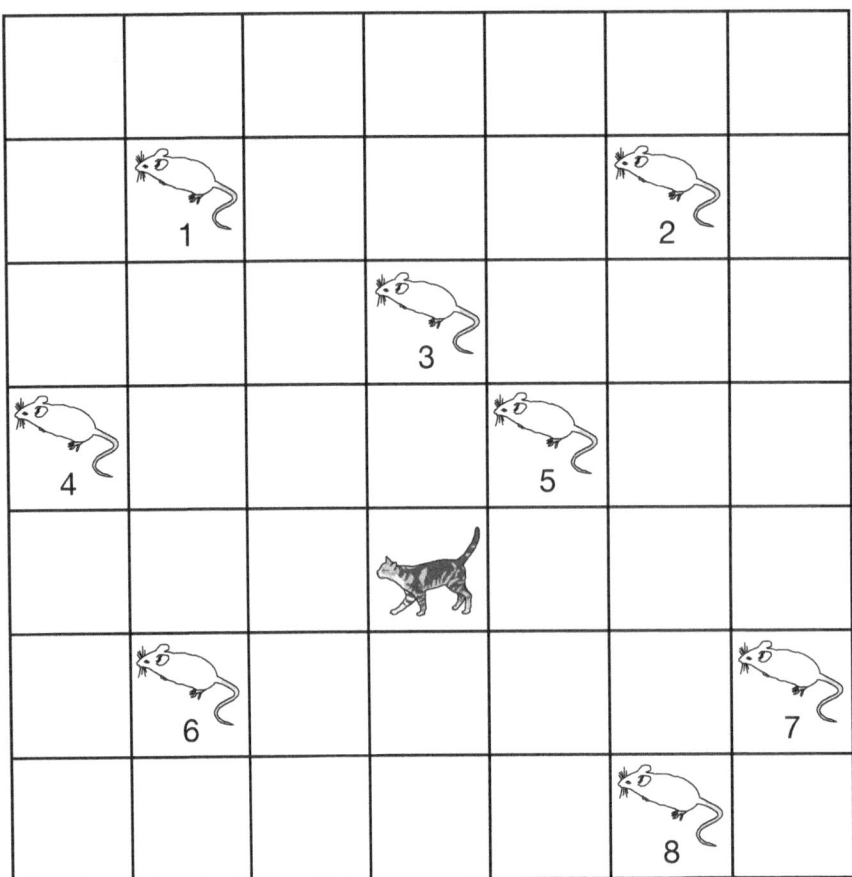

The cat moves 2 squares right and 3 squares up.

Write the number of the mouse that it catches.

2

1 mark

5 Write **one number** in each box to complete the calculations.

$$35 \div \boxed{10} = 3\cdot5$$

$$\boxed{650} \div 100 = 6\cdot5$$

1 mark

6 Write **all** the missing numbers in this multiplication grid.

×	3	7	9
2	6	14	18
7	21	49	63
11	33	77	99

2 marks

7 Draw an arrow (↓) to show the Roman numeral XL on the number line.

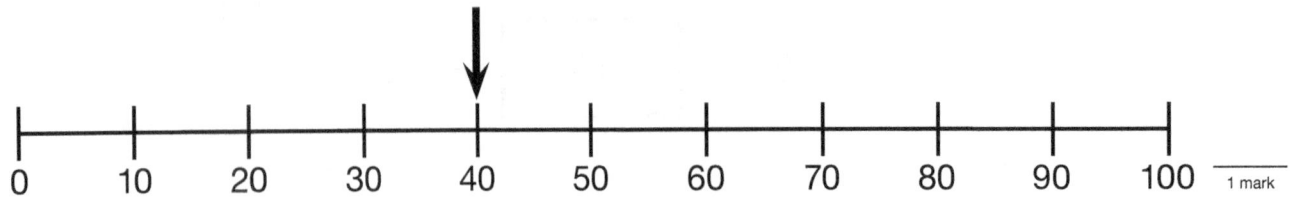

1 mark

8 Write the coordinates of the point marked with a cross (✗).

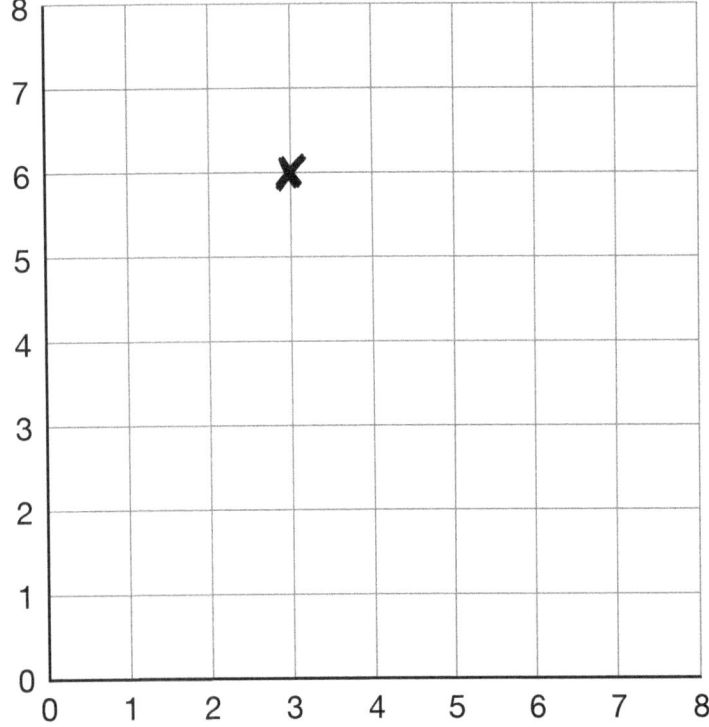

(3 , 6)

1 mark

9 There are 143 students in a school.

The Art department has £800 to spend.

They spend £5 per student on materials.

How much do they have left?

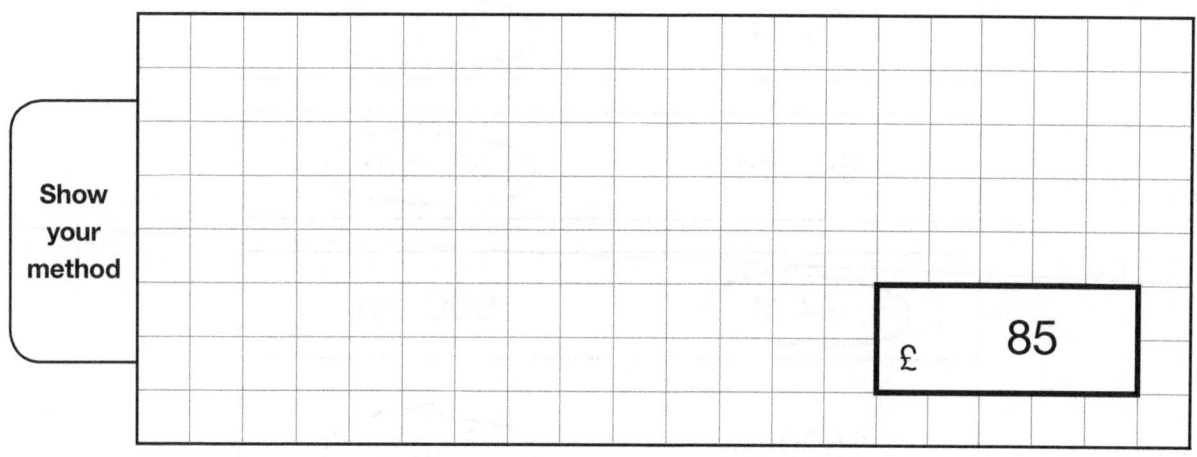

Show your method

£ 85

2 marks

10 Write a number in each box to make a correct calculation.

$$\boxed{2} \times \boxed{5} \times \boxed{10} = 100$$

1 mark

11 Circle the **larger** measure in each pair.

One has been done for you.

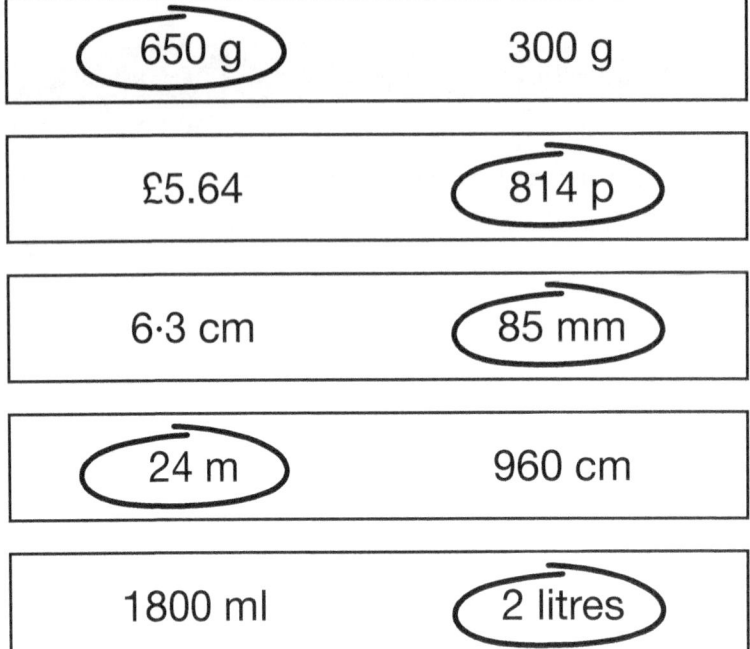

(650 g)	300 g
£5.64	(814 p)
6·3 cm	(85 mm)
(24 m)	960 cm
1800 ml	(2 litres)

2 marks

12 The graph shows the temperature in °C at different times in one day.

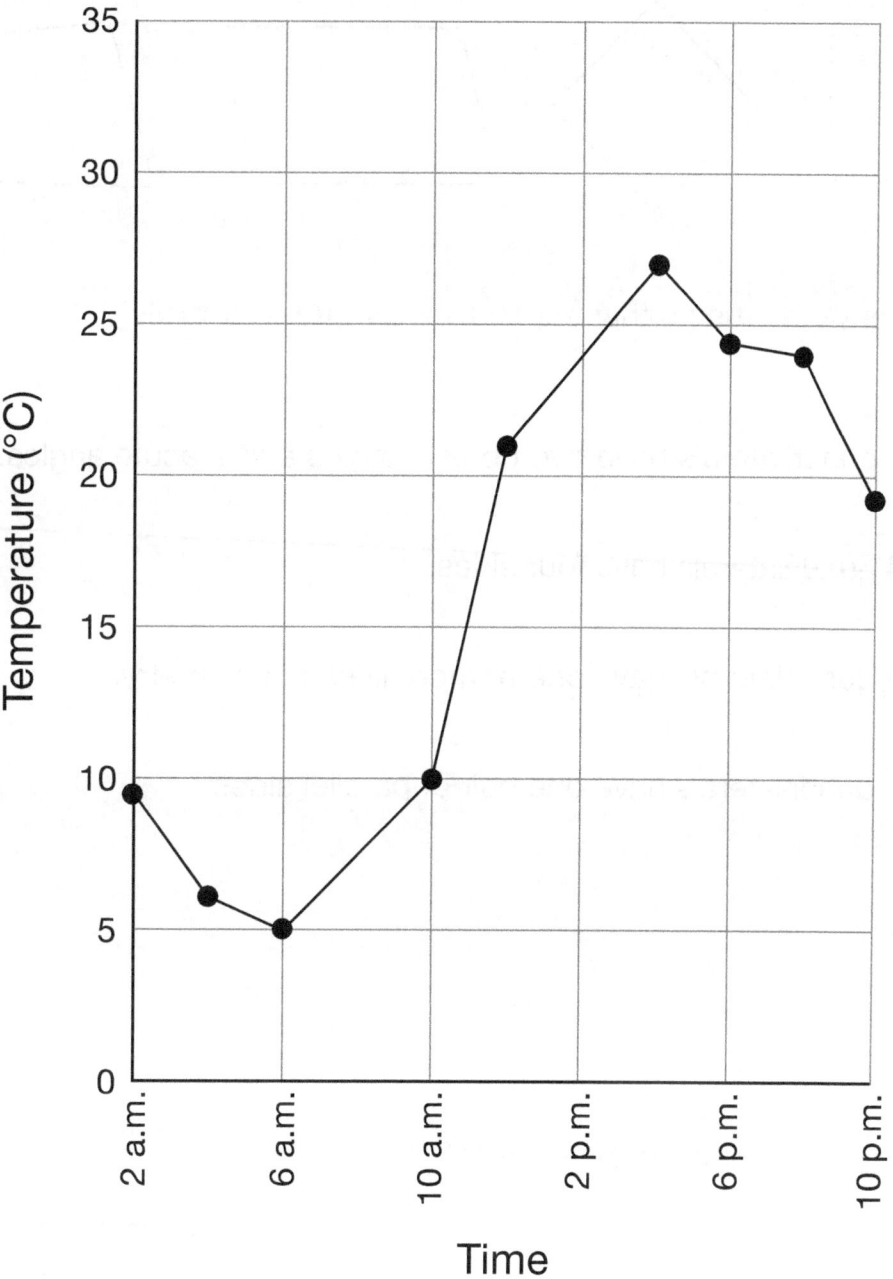

What was the temperature at 10 a.m.?

| 10 | °C |

Between which hours of the day was the temperature increasing?

6 a.m. and 4 p.m.

13 Some quadrilaterals are shown below.

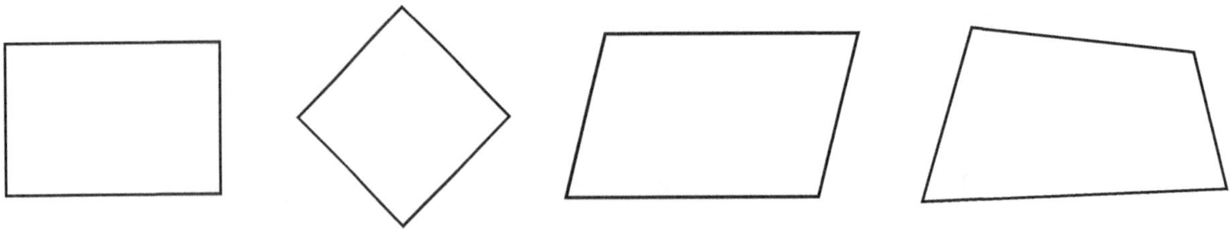

Tick (✔) any statements that are true for **all quadrilaterals**.

[ ] All quadrilaterals have two obtuse angles and 2 acute angles.

[✔] All quadrilaterals have four sides.

[ ] All quadrilaterals have one or more lines of symmetry.

[ ] All quadrilaterals have one pair of parallel sides.

1 mark

14 Write a number that is between 6·46 and 6·64

6·47

1 mark

15 A trapezium is made from three equilateral triangles.

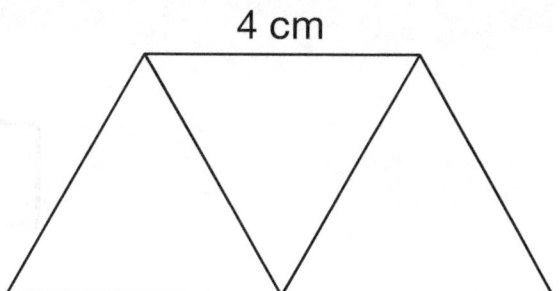

4 cm

What is the perimeter of the trapezium?

20   cm

1 mark

16 Circle the number that is a multiple of 4 **and** a multiple of 6

40	41	42	43
44	45	46	47
48	49	50	51
52	53	54	55

1 mark

17 Jonas measures the length of a classroom as $4\frac{3}{10}$ metres.

Write $4\frac{3}{10}$ as a decimal.

4·3

1 mark

18 There are 7 people in a netball team.

105 children sign up for a netball competition.

How many teams can be made?

15

1 mark

19 $\frac{1}{4}$ of the shape is shaded.

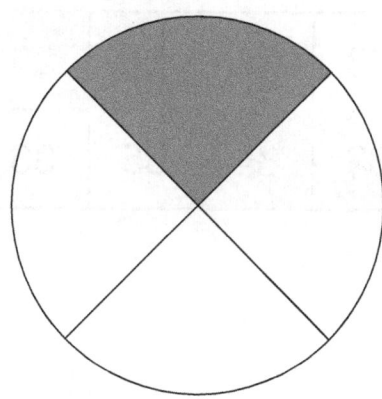

Write two fractions equivalent to $\frac{1}{4}$

$\boxed{\dfrac{2}{8}}$ and $\boxed{\dfrac{3}{12}}$

2 marks

20 Draw a rectangle with an area of 15 square centimetres.

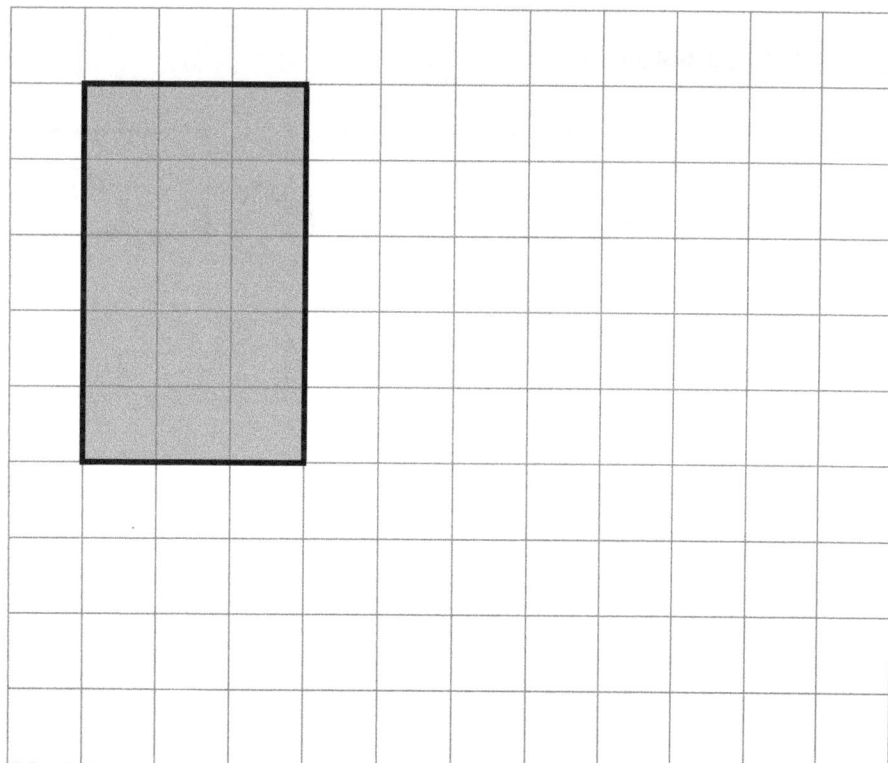

1 mark

21 Shade **all** the numbers that are 30 when rounded to the nearest 10

20	21	22	23	24	25	26	27	28	29
30	31	32	33	34	35	36	37	38	39

1 mark

22 Two numbers have a difference of 185

One of the numbers is 4560

Find the other **two** possible numbers.

| 4375 | and | 4745 |

2 marks

23 Write the missing numbers in the boxes.

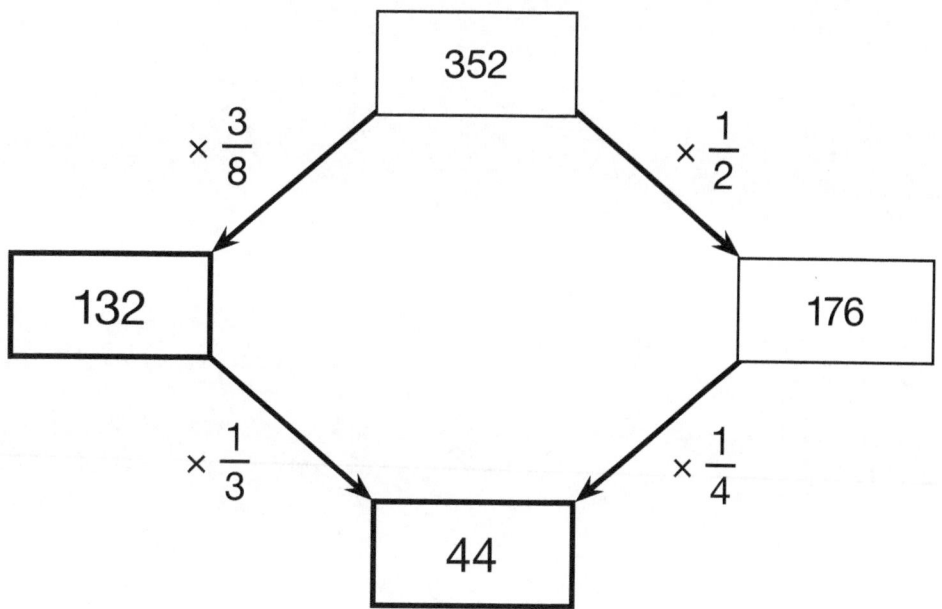

2 marks